Doz. Dr. Günther Windisch

Born in 1946 at Freiberg, Vogtland. Studied Mathematics from 1966 - 1970 at the Technical University of Karl-Marx-Stadt and receives Dr. rer.nat. in 1973, Dr. sc.nat. in 1985. Since 1988 Senior Lecturer in Numerical Mathematics at the Technical University of Karl-Marx-Stadt.

Fields of research: Stability of numerical methods, discretization methods for differential equations, primarily for parabolic problems, application of discretization methods in mathematics and engineering.

Windisch, Günther:
M-matrices in numerical analysis / Günther Windisch. -
1. Aufl.
(Teubner-Texte zur Mathematik ; 115)
NE: GT
ISBN 978-3-663-10819-1   ISBN 978-3-663-10818-4 (eBook)
DOI 10.1007/978-3-663-10818-4

1. Auflage
VLN 294-375/68/89 · LSV 1035
Gesamtherstellung: Druckerei "Magnus Poser" Jena
Betrieb des Graphischen Großbetriebes INTERDRUCK Leipzig
Bestell-Nr. 666 540 3
01750

Günther Windisch

M-matrices in Numerical Analysis

In the opening part this book gives a general survey of the theory of M-matrices, and in its main part, establishes a fairly close relationship between nonsingular M-matrices and discretization methods for second-order linear elliptic and parabolic problems. This approach applies to inverse-monotonicity, nonnegativity and monotonicity of solutions, maximum principles and conservation laws. It is shown how such properties carry over from continuous problems into their discrete approximations via nonsingular M-matrices, arising in the application of finite difference methods, finite element methods or the method of lines.

Im Einleitungsteil enthält das Buch einen allgemeinen Überblick über
die Theorie der M-Matrizen und vermittelt im Hauptteil enge Beziehun-
gen zwischen nichtsingulären M-Matrizen und Diskretisierungsmethoden
für lineare elliptische und parabolische Probleme. Dies betrifft die
Inversmonotonie, die Nichtnegativität und Monotonie von Lösungen so-
wie Maximumprinzipien und Erhaltungssätze. Es wird gezeigt, wie sol-
che Eigenschaften von kontinuierlichen Problemen in diskrete Approxi-
mationen durch nichtsinguläre M-Matrizen, die in der Anwendung von
Differenzenmethoden, der Methode der finiten Elemente oder der Linien-
methode entstehen, übertragen werden.

Dans la première partie, ce livre donne un aperçu général sur la théo-
rie des M-matrices, et, dans la partie principale, établir des relations étroites entre les M-matrices non singulières et les méthodes
de discrétisation entre les problèmes linéaires elliptiques et para-
boliques du second ordre cette approche s'applique à la monotonie inverse, à la non-négativité et à la monotonie des solutions ainsi que
aux principes du maximum et aux lois des conservation. On montre comment de telles propriétés peuvent être transposées du cas des problèmes
continus a celui de leurs approximations discrètes via les M-matrices
non singulières prenant naissance dans l'application des méthodes des
différences finies, des l'méthodes des éléments finis ou de la méthode
des lignes.

Монография содержит введение в общую теорию М-матриц и в её главной
части устанавливает тесную взаимосвязь между невырожденными М-матри-
цами и численными методами для решения линейных эллиптических и
параболических задач второго порядка. Этот подход включает понятия
обратной монотоности, неотрицательности и монотоности решений, а также
принцип максимума и законы сохранения. Показывается, как эти свойства
переносятся с непрерывных задач на дискретные приближения с помощью
М-матриц, возникающих при применении разностных методов, метода
конечных элементов или метода линий.

P R E F A C E

Many problems arising in the mathematical, physical, chemical, biological and social sciences lead to linear equation systems with some special system matrices A. This book is devoted to M-matrices, a subclass of matrices which can be expressed in the form $A = sI-B$, $s > 0$, $B \geqslant 0$ and I unit matrix. In recent years a great deal of research in M-matrices has been carried out and meanwhile many applications of M-matrices in different fields are known.

Our concern is the role of M-matrices in Numerical Analysis. Motivated by our own ten-year experience in discretization methods for solving numerically problems in mathematics and engineering, I undertook to write this book for two reasons. First, I intended to make the basics of M-matrix theory easily accessible. Second, I felt that there was an intimate relationship between nonsingular M-matrices and the discretization of second order linear elliptic and parabolic problems by finite difference methods, finite element methods and the method of lines.

Therefore this book is divided into two main parts. In the opening part, Chapters 2 and 3, the class of M-matrices is introduced and important M-matrix properties are discussed with an eye towards subsequent application in discretization methods. The purpose of the second part, Chapters 4, 5 and 6, is to show how basic properties of the continuous problems mentioned are reflected in discrete approximations by nonsingular M-matrices. One-dimensional problems are usually introductory while two-dimensional problems stand for the higher-dimensional case.

Certain special topics are not included for various reasons. Almost no mention is made of developments in singular M-matrix theory. Further, all investigations concerning the convergence of discretizations under consideration are excluded. Nonlinear problems are also omitted from the discussion.

I am grateful to Prof.V.Friedrich for being a constant source of inspiration and encouragement over the years. I wish to thank also Mrs. M.Pester for help in preparing the text. Further, I would like to thank Dr.phil.B.Legler for revising the English version of this book. Moreover, I feel obliged to Dr.R.Müller and to Mrs. Roth of Teubner-Verlag for the permanent support throughout the preparation of the manuscript.

Karl-Marx-Stadt, April 1989

Günther Windisch

C O N T E N T S

N O T A T I O N

List of frequently used symbols

n	natural number
N	$N = \{1,\dots,n\}$
$x,y,..$	n-dimensional column vectors, $x = (x_i)$
$x \leqslant y$	natural partial ordering: $x_i \leqslant y_i$, $\forall i \in N$
$x < y$	$x_i < y_i$, $\forall i \in N$
$A,B,..$	n×n matrices, $A = (a_{ij})$
$A \leqslant B$	natural partial ordering: $a_{ij} \leqslant b_{ij}$, $\forall i,j \in N$
$A < B$	$a_{ij} < b_{ij}$, $\forall i,j \in N$
$(Ax)_i$	i-th component of Ax
A^T, x^T	transpose of A, x
A^{-1}	inverse of A
A^k	$A^k = (a_{ij}^{(k)})$, k-th power of A for integers k
det A	determinant of A
tr A	trace of A
diag A	$= \mathrm{diag}(a_{11},\dots,a_{nn})$
$\sigma(A)$	spectrum of A
$S(A)$	spectral radius of A
adj A	adjoint of A
$\mathcal{G}(A)$	associated directed graph of A
$Z^{n \times n}$	$= \{A = (a_{ij})$ of order n, $a_{ij} \leqslant 0$ for $i \neq j\}$
I	unit matrix
D	positive diagonal matrix, $D = \mathrm{diag}(d_1,\dots,d_n)$, $d_i > 0$, $\forall i \in N$
P	permutation matrix
e	the vector of all ones, $e = (1,\dots,1)^T$
e_i	i-th coordinate unit vector
$N_+(x)$	$= \{i \in N: \; x_i > 0\}$
$N_-(x)$	$= \{i \in N: \; x_i < 0\}$
$N_0(x)$	$= \{i \in N: \; x_i = 0\}$
$|x|, |A|$	$|x| = (|x_i|)$, $|A| = (|a_{ij}|)$

$\|x\|$, $\|x\|_p$ vector norm of x, l_p-norm of x for $p \geqslant 1$

$\|A\|$, $\|A\|_p$ matrix norm of A, l_p-norm of A for $p \geqslant 1$

R^1 real numbers

R^n real n-dimensional linear space of column vectors

R_+^n nonnegative orthant

$\mathbb{C}$ complex numbers

$\mathcal{K}_r(z_0) = \{ z \in \mathbb{C} : \ |z - z_0| < r \}$, circle in $\mathbb{C}$

Ω bounded region in R^n

$\partial\Omega$ boundary of Ω

$\overline{\Omega} = \Omega + \partial\Omega$, closure of Ω

ν outward normal to Ω

$C^k(\Omega)$ k-times continuously differentiable functions in Ω

$L_2(\Omega)$ space of measurable functions whose 2-th power is Lebesgue integrable

$W_2^k(\Omega)$ Sobolev space $W_2^k(\Omega)$

$$\nabla u = \left(\frac{\partial u}{\partial x_1}, \ldots, \frac{\partial u}{\partial x_n} \right)^T, \quad \Delta u = \sum_{i=1}^{n} \frac{\partial^2 u}{\partial x_i^2}$$

(u,v) scalar product

$a(u,v)$ bilinear form

h, τ discretization parameters

$\overline{\omega}_h = \omega_h + \gamma_h, \quad \omega_h \subset \Omega, \quad \gamma_h \subset \partial\Omega$

D_+, D_-, D_o difference operators

$$D_+ y_i = \frac{y_{i+1} - y_i}{h}, \quad D_- y_i = \frac{y_i - y_{i-1}}{h},$$

$$D_o y_i = \frac{y_{i+1} - y_{i-1}}{2h}, \quad D_+ D_- y_i = \frac{y_{i-1} - 2y_i + y_{i+1}}{h^2},$$

$O(h), o(h)$ Landau symbols

$\mathcal{S}(i)$ approximation star

FDM   finite difference method

FEM   finite element method

ML   method of lines

1. DEFINITIONS AND PROPOSITIONS

The opening chapter is devoted to an introduction to some basic
definitions required in our later considerations. Furthermore, we
assemble a variety of propositions from the literature, which will
be needed in the sequel. The cited results are classical and are
found in most standard books on the subject. For the proofs we refer
to the literature.

We assume that all matrices under consideration are real and square
of order $n \geqslant 2$ unless specified otherwise.

<u>Definition 1.1.</u> (diagonally dominant matrices,[2,18,28])

A matrix $A = (a_{ij})$ is said to be strongly row diagonally dominant,
if

$$|a_{ii}| > \sum_{j \neq i} |a_{ij}| \, , \quad \forall i \in N.$$

A matrix $A = (a_{ij})$ is said to be weakly row diagonally dominant,
if

$$|a_{ii}| \geqslant \sum_{j \neq i} |a_{ij}| \, , \quad \forall i \in N,$$

and, for at least one $i \in N$, the strong inequality holds.

If A^T is strongly (weakly) row diagonally dominant, then A is said
to be strongly (weakly) column diagonally dominant.

<u>Proposition 1.2.</u>, [18]

Let A be a strongly row or column diagonally dominant matrix.
Then, $\det A \neq 0$.

<u>Proposition 1.3.</u>, [28]

Let $A = (a_{ij})$ be a strongly row diagonally dominant matrix with

$$r_i = |a_{ii}| - \sum_{j \neq i} |a_{ij}| > 0, \quad \forall i \in N.$$

Then

$$\| A^{-1} \|_\infty \leqslant \max_{i \in N} \frac{1}{r_i} \, .$$

The bound is strict. Equality holds, for instance, for all strongly
diagonally dominant diagonal matrices $A = \mathrm{diag}(a_{11}, \ldots, a_{nn})$.

<u>Definition 1.4.</u> (entry diagonally dominant matrices,[40])

A matrix $A = (a_{ij})$ is said to be strongly diagonally dominant of its row entries, if for all $i \in N$

$$|a_{ii}| > |a_{ij}| , \quad \forall j \neq i,$$

holds.

If $|a_{ii}| \geqslant |a_{ij}|$, $\forall j \neq i$, $\forall i \in N$, then A is said to be weakly diagonally dominant of its row entries.

A matrix A is said to be strongly (weakly) diagonally dominant of its column entries, if A^T is strongly (weakly) diagonally dominant of its row entries.

<u>Proposition 1.5.</u>

Let A be a strongly (weakly) row (column) diagonally dominant matrix. Then A is strongly (weakly) diagonally dominant of its row (column) entries.

<u>Definition 1.6.</u> (reducible, irreducible matrices,[2,10,18,28])

A matrix A of order $n \geqslant 2$ is called reducible, if

$$P^T A P = \begin{pmatrix} A_{11} & A_{12} \\ 0 & A_{22} \end{pmatrix},$$

for some permutation matrix P, with A_{11} and A_{22} square.
A matrix A is called irreducible if it is not reducible.

<u>Proposition 1.7.</u>

Let A be a reducible matrix. Then A^k is reducible for each integer $k \geqslant 2$. Let A be nonsingular and reducible, then A^{-k} is reducible for each integer $k \geqslant 1$.

<u>Proposition 1.8.</u> ,[18]

Let $A = (a_{ij})$ be an irreducible weakly row or column diagonally dominant matrix. Then $\det A \neq 0$ and $a_{ii} \neq 0$, $\forall i \in N$.

<u>Proposition 1.9.</u> ,[10]

Let $A \geqslant 0$ be an irreducible matrix and let $B > 0$, both matrices of the same order. Then $AB > 0$ and $BA > 0$.

<u>Proposition 1.10.</u> ,[10]

Let $A = (a_{ij}) \geqslant 0$ be irreducible of order n. Then $(I + A)^{n-1} > 0$ and, if additionally all diagonal entries of A are positive,

9

even $A^{n-1} > 0$.

Denote $A^k = (a_{ij}^{(k)})$ for integers $k \geqslant 1$. Then for every pair (i,j) there exists an integer $k \leqslant m$, m the degree of the minimal polynomial of A, such that $a_{ij}^{(k)} > 0$.

Definition 1.11. (associated directed graph of a matrix,[2,18,29])

The associated directed graph $\mathcal{G}(A)$ of a matrix $A = (a_{ij})$ of order n consists of n vertices $P_1,\ldots,P_n$ where an edge leads from P_i to P_j if and only if $a_{ij} \neq 0$, $i \neq j$.
A directed graph $\mathcal{G}(A)$ is strongly connected if for any ordered pair P_i,P_j of vertices of $\mathcal{G}(A)$ there exists a sequence of edges (a path) which leads from P_i to P_j.

Proposition 1.12. ,[2]

A matrix A is irreducible if and only if the associated directed graph $\mathcal{G}(A)$ is strongly connected.

Proposition 1.13.

Let A be an irreducible matrix. Then the matrices $-A$ and A^T are also irreducible.
Let $A \geqslant 0$ be irreducible and let $B \geqslant 0$, both matrices of the same order. Then $A+B$ is irreducible.

Definition 1.14. (definite matrices,[2,10,18])

A matrix A is called positive definite if
$$x^T A x > 0, \quad \forall x \in R^n, \quad x \neq 0.$$

If A satisfies
$$x^T A x \geqslant 0, \quad \forall x \in R^n,$$

then A is called positive semidefinite.

Proposition 1.15. ,[18]

Let $A = A^T = (a_{ij})$ be strongly diagonally dominant with $a_{ii} > 0$, $\forall i \in N$. Then A is positive definite.

Proposition 1.16. ,[18]

Let $A = A^T = (a_{ij})$ be irreducible and weakly diagonally dominant with $a_{ii} > 0$, $\forall i \in N$. Then A is positive definite.

Definition 1.17. (spectral radius of matrices,[2,18,21])

Let $\sigma(A) = \{\lambda_i\}_{i \in N}$ be the spectrum of A.
Then
$$S(A) = \max_{i \in N} |\lambda_i| \ ,$$

is called the spectral radius of the matrix A.

Proposition 1.18. ,[21]

Let $B \geqslant 0$, $C = (c_{ij})$, $c_{ij} \in \mathbb{C}$ for $\forall i,j \in N$, where $|C| = (|c_{ij}|) \leqslant B$.
Then
$$S(C) \leqslant S(B) \ .$$

Proposition 1.19. ,[19,28,29]

For any matrix norm $\|A\|$ with $\|Ax\| \leqslant \|A\| \ \|x\|$, $\forall x \in R^n$, there holds
$$S(A) \leqslant \|A\| \ .$$

Thus, $\|A\| < 1$ implies $S(A) < 1$.
Let A be irreducible with $\|A\|_\infty = 1$, and, for at least one $i \in N$, we
assume
$$\sum_{j=1}^{n} |a_{ij}| < 1 \ .$$

Then $S(A) < 1$.

Proposition 1.20. ,[19]

The spectrum $\sigma(A)$ of any matrix $A = (a_{ij})$ is enclosed in the union
of the Gershgorin disks
$$\sigma(A) \subset \bigcup_{i \in N} \left\{ z \in \mathbb{C} : |z - a_{ii}| \leqslant \sum_{j \neq i} |a_{ij}| \right\} \ .$$

Definition 1.21. (convergent, semiconvergent matrices,[29,30])

A matrix A is called convergent if
$$\lim_{k \to \infty} A^k = 0 \ ,$$

and semiconvergent if
$$\lim_{k \to \infty} A^k \text{ exists.}$$

Proposition 1.22. ,[29]

A matrix A is convergent if and only if $S(A) < 1$.

Proposition 1.23. ,[30]

A matrix A is semiconvergent if and only if each of the following

conditions holds
  (a) $S(A) \leq 1$,
  (b) if $S(A) = 1$, then $\lambda \in \sigma(A)$ with $|\lambda| = 1$ implies $\lambda = 1$,
  (c) if $S(A) = 1$, then all elementary divisors associated with the
       eigenvalue 1 of A are linear.

<u>Definition 1.24.</u> (Neumann series of A,[10,19])

The infinite sum

$$R(\lambda) = \sum_{k=0}^{\infty} \lambda^{-(k+1)} A^k,$$

is called the Neumann series of $A, \lambda \in R^1$.

<u>Proposition 1.25.</u> ,[19]

The Neumann series of A converges if and only if $S(A) < |\lambda|$ and it
follows that $R(\lambda) = (\lambda I - A)^{-1}$.

<u>Proposition 1.26.</u> ,[10]

If $A \geq 0$, then there exists $(\lambda I - A)^{-1}$ for all $\lambda > S(A)$ and
$(\lambda I - A)^{-1} \geq 0$.

<u>Definition 1.27.</u> (Perron-Frobenius eigenvalue of $A \geq 0$,[2,10,28])

Let $A \geq 0$, then $S(A)$ is called the Perron-Frobenius eigenvalue
of A.

<u>Proposition 1.28.</u> ,[10]

Let $A \geq 0$. Then the spectral radius $S(A)$ is an eigenvalue of A with
at least one eigenvector $x \geq 0$.

<u>Proposition 1.29.</u> ,[10]

For an irreducible matrix $A \geq 0$, there cannot exist two linear
independent nonnegative eigenvectors.

<u>Proposition 1.30.</u> , Perron theorem ,[10]

If $A > 0$, then $S(A)$ is the only eigenvalue of A of modulus $S(A)$. The
corresponding eigenvector x, i.e. $Ax = S(A)x$, can be chosen such
that $x > 0$.

<u>Proposition 1.31.</u> , Perron-Frobenius theorem ,[10]

Let $A \geq 0$ be irreducible. Then the following holds
  (a) The spectral radius $S(A)$ is a simple eigenvalue of A and the
       corresponding eigenvector x can be chosen such that $x > 0$.

(b) If A has k eigenvalues of modulus S(A), then these numbers are
distinct roots of the equation

$$\lambda^k - (S(A))^k = 0.$$

(c) The whole spectrum $\sigma(A)$ goes over into itself under a rotation
of the complex plane by the angle $2\pi/k$.

Proposition 1.32. ,[28]

Let $A = (a_{ij}) \geqq 0$ of order $n \geqq 2$ be irreducible with the spectral radius $S(A)$. Then

$$S(A) > \max_{i \in N} a_{ii}.$$

Proposition 1.33. ,[28]

Let $A \geqq 0$ and be denoted by s and t, the smallest and greatest row sums of A, respectively. Then

$$s \leqq S(A) \leqq t,$$

with equality on either side, implying equality throughout.

Proposition 1.34. ,[28]

If $A \geqq 0$ and $B \geqq 0$ are matrices of the same order, then

$$\max (S(A),S(B)) \leqq S(A + B).$$

Proposition 1.35. ,[28]

If $A = (a_{ij}) \geqq 0$ is irreducible, $|B| = (|b_{ij}|) \leqq A$ and $|b_{ij}| < a_{ij}$ for at least one pair (i,j) then

$$S(B) < S(A).$$

Proposition 1.36. ,[2]

Let A' be a principal submatrix of $A \geqq 0$ of order less than n. Then

$$S(A') \leqq S(A).$$

Much more precisely:
(a) If A is an irreducible matrix, then $S(A') < S(A)$.
(b) If A is a reducible matrix, then for at least one principal
submatrix A' of A, there holds

$$S(A') = S(A).$$

Definition 1.37. (primitive matrices,[2,26])

An irreducible matrix $A \geqq 0$ is called primitive if S(A) is the
only eigenvalue of A having modulus S(A).

Proposition 1.38. ,[2]

For an irreducible matrix $A \geqslant 0$, the following assertions are equivalent
  (a) A is primitive,
  (b) $\lim_{k \to \infty} (S(A)^{-1}A)^k$ exists,
  (c) there exists an integer $q \geqslant 1$ such that $A^q > 0$,
  (d) A^k is irreducible for all integers $k \geqslant 1$.

Proposition 1.39. ,[2]

If $A \geqslant 0$ is irreducible and $B \geqslant 0$ has tr $B > 0$, then $A + B$ is primitive.

Definition 1.40. (monotone matrices,[29])

A matrix A is called a monotone matrix if det $A \neq 0$ and $A^{-1} \geqslant 0$.

Proposition 1.41. ,[29]

A matrix A is a monotone matrix if and only if from $Ax \geqslant 0$ it follows that $x \geqslant 0$. Thus, $Ax \leqslant Ay$ implies $x \leqslant y$.

Proposition 1.42. ,[29]

The class of monotone matrices is closed under matrix multiplication. That is, if A and B are monotone matrices of the same order, then AB and BA are also monotone matrices.

Proposition 1.43. ,[29]

Let A be a monotone matrix and $f \geqslant 0$. Then the unique solution of the linear equation system $Ax = f$ is nonnegative, i.e. $x = A^{-1}f \geqslant 0$. Furthermore, let x_1 and x_2 be such that

$$Ax_1 \leqslant f = Ax \leqslant Ax_2,$$

holds, then the solution x of $Ax = f$ is enclosed by

$$x_1 \leqslant x \leqslant x_2.$$

Definition 1.44. (essentially positive matrices,[26])

A matrix $A = (a_{ij})$ is called essentially positive if A is irreducible and $a_{ij} \geqslant 0$, $i \neq j$.

Proposition 1.45. ,[26]

A matrix A is an essentially positive matrix if and only if $A + sI$ is a nonnegative, irreducible and primitive matrix for sufficiently large $s > 0$.

<u>Proposition 1.46.</u> ,[26]

A matrix A is essentially positive if and only if $\exp(tA) > 0$
for all $t > 0$.

<u>Proposition 1.47.</u> ,[26]

Let A be an essentially positive matrix. Then A has a real eigenvalue
λ_o such that

  (a)   to λ_o there corresponds an eigenvector $x > 0$,
  (b)   if λ is any other eigenvalue of A, then $\mathrm{Re}\,\lambda < \lambda_o$,
  (c) λ_o increases when any entry of A increases.

<u>Definition 1.48.</u> (regular splitting of a matrix,[29])

  A splitting $A = Q - R$ of a matrix A is called a regular splitting,
if Q is a monotone matrix and $R \geqslant 0$.
It is called a weak regular splitting if the condition $R \geqslant 0$ is
replaced by $Q^{-1}R \geqslant 0$ and $RQ^{-1} \geqslant 0$.

Any regular splitting of a matrix is a weak regular splitting.
The converse is not true.

<u>Proposition 1.49.</u> ,[29]

Let $A = Q - R$ be a regular splitting of a nonsingular matrix A.
Then A is a monotone matrix if and only if $S(Q^{-1}R) < 1$.

<u>Proposition 1.50.</u> ,[29]

Let A be a monotone matrix and $A = Q - R$ be a regular splitting
of A.
Then

$$S(Q^{-1}R) \;=\; \frac{S(A^{-1}R)}{1 + S(A^{-1}R)} \;<\; 1.$$

Lastly, we cite the Sherman-Morrison formula for rank-one perturbations of a nonsingular matrix , which will be needed in the sequel.

<u>Proposition 1.51.</u> ,[21,28]

Let A be a nonsingular matrix. Then $(A + uv^T)^{-1}$ exists if and only
if $1 + v^T A^{-1} u \neq 0$, and

$$(A + uv^T)^{-1} \;=\; A^{-1} \;-\; \frac{1}{1 + v^T A^{-1} u}\, A^{-1}uv^T A^{-1}\;.$$

2. M-MATRICES

The purpose of this chapter is to introduce the class of M-matrices.
The definition of M-matrices includes both singular and nonsingular
M-matrices, the former being only of particular interest to later
sections. We give a first variety of M-matrices to illustrate M-matrix properties in the next chapter. Further, we are concerned with
necessary and sufficient conditions for a matrix $A \in Z^{n \times n}$ to be an
M-matrix. Owing to some practical requirements, we are preferably
interested in sufficient M-matrix conditions, taking into account
only relations between the entries of a matrix.

2.1. Introduction to M-matrices

The term "M-matrix" (<u>M</u>inkowski <u>matrix</u>) was first used by A.Ostrowski
in [44] in reference to the work of Hermann Minkowski (distinguished
German mathematician, 1864 - 1907), who proved [42] , in our termi-
nology, that if a matrix $A \in Z^{n \times n}$ has all of its row sums positive,
then the determinant of A is positive, i.e.

$$A \in Z^{n \times n}, \quad Ae > 0 \qquad \text{implies} \qquad \det A > 0.$$

The first systematic effort to characterize M-matrices was made in
the earlier works of M.Fiedler and V.Pták,[34].
A large number of properties of M-matrices are known by now and have
found widespread application in many fields, see [2,9,12,21,26] and
elsewhere.

To begin with, let us accept the following general definition of
M-matrices.

<u>Definition 2.1.</u> (M-matrices,[2])

Any matrix A of the form

$$A = sI - B, \tag{2.1}$$

with $s > 0$, $B \geq 0$ for which $s \geq S(B)$, the Perron-Frobenius eigenvalue of B, is called an M-matrix.

Since $A = sI - B$, $B \geq 0$, it should come as no surprice that the
theory of nonnegative matrices plays a dominant role in the study
of M-matrices.
Obviously, the off-diagonal entries of any M-matrix $A = (a_{ij})$ are
nonpositive, i.e. $a_{ij} \leq 0$ for all $i \neq j$. Thus, the class of M-matrices is a subset of $Z^{n \times n}$.

For $s = S(B)$, the matrix A in the representation (2.1) is a singular
M-matrix. By Proposition 1.28., there exists an eigenvector $x \geqslant 0$ of B
such that $Bx = S(B)x$. Thus $Ax = (S(B)I - B)x = 0$, which shows $0 \in \sigma(A)$
and, therefore, det $A = 0$.

For $s > S(B)$, the M-matrices are nonsingular.

The following theorem characterizes nonsingular M-matrices.

<u>Theorem 2.2.</u>

For $s > S(B)$, any M-matrix $A = sI - B$ is a monotone matrix.

Proof. The assertion of Theorem 2.2. is an immediate consequence of
Proposition 1.26. □

It should be mentioned that M-matrices are introduced in earlier
works as nonsingular matrices by the following definition, see [21,29]
and elsewhere.

<u>Definition 2.3.</u> (nonsingular M-matrices, [21])

A matrix $A = (a_{ij})$ is called a nonsingular M-matrix if det $A \neq 0$,
$A^{-1} \geq 0$ and $a_{ij} \leq 0$ for all $i \neq j$.

We infer that Definition 2.3. is equivalent to Definition 2.1. for
$s > S(B)$.
For the proof, we assume $s > S(B)$ in Definition 2.1. By Theorem 2.2.
and by the comment that $a_{ij} \leq 0$ for $i \neq j$ in (2.1), it follows that
Definition 2.3. is a conclusion of Definition 2.1.
Conversely, let $A = (a_{ij})$ be a nonsingular M-matrix by Definition 2.3.
For $s > \max \{0, \max_{i \in N} a_{ii}\}$ let $A = sI - B$, where $B = sI - A \geq 0$. Then
$A = sI - B$ is a regular splitting of the monotone matrix A and it
follows from Proposition 1.49. that $S(B/s) = S(B)/s < 1$. Thus,
$A = sI - B$ is of the form (2.1) with $s > S(B)$. This completes the
proof. □

We now state a theorem on the sign of the diagonal entries of nonsingular M-matrices and its inverse.

<u>Theorem 2.4.</u> , [29]

Any nonsingular M-matrix $A = (a_{ij})$ and its inverse $A^{-1} = (a_{ij}^-)$ have
all positive diagonal entries.

Proof. From the identity $AA^{-1} = I$ it follows that

$$\sum_{k \in N} a_{ik} a_{ki}^- = 1 , \quad \forall i \in N.$$

By $a_{ij} \leq 0$, $i \neq j$ and $A^{-1} \geq 0$, we have

$$a_{ii} a_{ii}^- = 1 - \sum_{k \neq i} a_{ik} a_{ki}^- \geq 1.$$

Thus, $a_{ii}^- > 0$ instead of $a_{ii}^- \geq 0$, and additionally $a_{ii} > 0$, $\forall i \in N$. The proof is complete.

□

Exploiting that $A = sI - B$ is a binomial of $B \geq 0$, we can apply the spectral theory for nonnegative matrices to investigate some spectral properties of M-matrices.

<u>Theorem 2.5.</u> ,[2]

Let $A = sI - B$ be an <u>M-matrix</u>. Then

$$\sigma(A) \subset \overline{\mathcal{K}_{S(B)}(s)} \subset \left\{ z \in \mathbb{C} : \operatorname{Re} z \geq 0 \right\}.$$

Proof. By Proposition 1.28., we have $\sigma(B) \subset \overline{\mathcal{K}_{S(B)}(0)}$. Thus, any $\lambda \in \sigma(A)$ has the representation $\lambda = s - \mu$, $\mu \in \sigma(B)$. Our assertion follows then from $s \geq S(B)$.

□

This last result shows that
  (a) if A is a nonsingular M-matrix, i.e. $s > S(B)$, then $\operatorname{Re} \lambda > 0$
      for any $\lambda \in \sigma(A)$,
  (b) if A is a singular M-matrix, i.e. $s = S(B)$, then $\operatorname{Re} \lambda > 0$ for
      any $\lambda \in \sigma(A) \setminus \{0\}$.

It is easily seen by Proposition 1.13. that a matrix A of the form $A = sI - B$ is irreducible if and only if B is irreducible.

For irreducible M-matrices we shall state the next theorem.

<u>Theorem 2.6.</u> ,[2]

Any irreducible M-matrix A has a simple real eigenvalue of smallest modulo and the corresponding eigenvector x can be chosen such that $x > 0$.

Proof. Let $A = sI - B$ be an irreducible M-matrix, which implies that $B \geq 0$ is irreducible. By Proposition 1.31., $S(B)$ is a simple eigenvalue of B and the corresponding eigenvector x can be chosen such that $x > 0$. Thus, $\lambda = s - S(B)$ is a simple real eigenvalue of A and it is of smallest modulo. Furthermore, the eigenvector $x > 0$ corresponds to the eigenvalue $\lambda = s - S(B) \in \sigma(A)$.

□

Thus, Theorem 2.6. yields the following corollary.

<u>Corollary 2.7.</u>

For an irreducible M-matrix $A = sI - B$, there cannot exist two linear independent nonnegative eigenvectors.

Proof. Since any eigenvector of $B \geq 0$ is also an eigenvector of
$A = sI - B$, the assertion follows from Proposition 1.29.

Next, let us illustrate the positions of the spectrum $\sigma(A)$ of
M-matrices A in the complex plane geometrically.
We have $\sigma(A) = \overline{\sigma(A)}$, because any M-matrix is a real matrix.

Fig.2.1.   • :   points of $\sigma(A)$

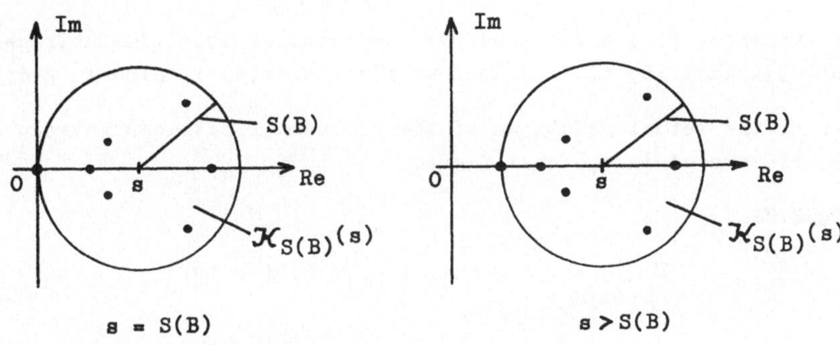

Let us take an example of an M-matrix A to demonstrate the case
where
$$\sigma(A) \subset \partial \mathcal{K}_{S(B)}(s) = \left\{ z \in \mathbb{C} : |z - s| = S(B) \right\}.$$

We choose
$$B = \begin{pmatrix} 0 & 1 & & O \\ & & \ddots & \\ & & & 1 \\ 1 & O & & 0 \end{pmatrix},$$

which is a permutation matrix. The spectrum $\sigma(B)$ is defined by the
roots of the characteristic polynomial
$$\det(\mu I - B) = \mu^n - 1 = 0.$$

We have $S(B) = 1$ and for $s \geq 1$ the matrix $A = sI - B$ is an M-matrix
with $\sigma(A) \subset \partial \mathcal{K}_1(s)$.

Let us now turn to the symmetric M-matrices.

Definition 2.8. (Stieltjes matrices,[29])

   A symmetric nonsingular M-matrix A is called a Stieltjes
   matrix.

We remark that a nonsingular M-matrix $A = sI - B$ is a Stieltjes
matrix if and only if $B = B^T$. Any Stieltjes matrix A is positive

definite, i.e. $\sigma(A) \subset R_+^1 \setminus \{0\}$.

Our next object is to introduce the class of L-matrices.

<u>Definition 2.9.</u> (L-matrices,[29])

A matrix $A = (a_{ij})$ is called an L-matrix if $a_{ii} > 0$, $\forall i \in N$
and $a_{ij} \leq 0$, $i \neq j$.

It is then clear from Theorem 2.4. that any nonsingular M-matrix is
an L-matrix. The converse is not true.

For functions f(t) which preserve the class of Stieltjes matrices,
see [43]. That is, if A is an L-matrix then f(A) is also an L-matrix.

Some of the mutual relations of the classes of matrices introduced
are illustrated in the next figure.

Fig.2.2.

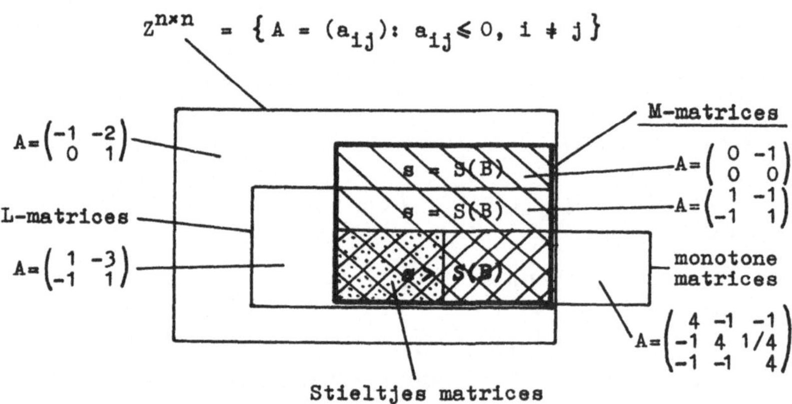

We conclude this section with a further definition.

<u>Definition 2.10.</u> (M-matrices with "property c",[30])

An M-matrix A = sI - B is said to have "property c" if the
matrix T = B/s is semiconvergent.

All nonsingular M-matrices have "property c" because of $S(T) = S(B)/s < 1$, which shows that T is convergent, see Proposition 1.22.
Furthermore, by Proposition 1.38. it is seen that a singular M-matrix
A = S(B)I - B has "property c" if $B \geq 0$ is primitive.
There are, however, singular M-matrices A which do not share

"property c", see for example the matrix

$$A = \begin{pmatrix} 0 & -1 \\ 0 & 0 \end{pmatrix} .$$

We have for any $s > 0$

$$A = sI - B , \quad B = \begin{pmatrix} s & 1 \\ 0 & s \end{pmatrix} , \quad T = \begin{pmatrix} 1 & 1/s \\ 0 & 1 \end{pmatrix} .$$

Taking the limit of

$$T^k = \begin{pmatrix} 1 & k/s \\ 0 & 1 \end{pmatrix},$$

for $k \rightarrow \infty$, T is not semiconvergent.

2.2. Examples of M-matrices

We are now going to illustrate some types of M-matrices. We start
with examples which are direct perturbations of the unit matrix I.

2.2.1. Rank-one perturbations of the unit matrix

Let $u \geqslant 0$, $v \geqslant 0$ be n-dimensional column vectors. We consider the
rank-one perturbation of the unit matrix of the form

$$A = I - r u v^T , \tag{2.2}$$

where $r > 0$ is a constant. On comparison with (2.1), we have $s = 1$
and $B = ruv^T \geqslant 0$. We state

<u>Lemma 2.11.</u>

The matrix $A = I - ruv^T$ with $u \geqslant 0$, $v \geqslant 0$, $r > 0$ is an M-matrix if
and only if

$$r v^T u \leqslant 1 . \tag{2.3}$$

For $rv^T u < 1$, the matrix A is a nonsingular M-matrix, for $rv^T u = 1$
it is a singular one.

Proof. We first assume that (2.3) holds. From

$$\mathfrak{S}(ruv^T) = \{0, rv^T u\}, \tag{2.4}$$

and (2.3) it follows that $S(ruv^T) = rv^T u \leqslant 1 = s$, which shows that
A is an M-matrix. The matrix A is nonsingular for $rv^T u < 1$ and sin-
gular for $rv^T u = 1$.
Next, let $A = I - ruv^T$ be a nonsingular M-matrix. In this case,
A is also monotone by Theorem 2.2., and (2.2) is a regular splitting
of A. Thus, by Proposition 1.49., $S(ruv^T) < 1$, which proves $rv^T u < 1$.
If A is a singular M-matrix, then $rv^T u = 1$ which completes the proof.
$\square$

We remark that $A = I - ruv^T$ is irreducible if and only if all the
components of $u=(u_i)$ and $v=(v_i)$ are nonzero, i.e. if $\prod_{i=1}^{n} u_i v_i \neq 0$.

In our case this means that A is irreducible if and only if $u > 0$ and $v > 0$. Then $S(ruv^T) = rv^Tu > 0$, which is the simple Perron-Frobenius eigenvalue of $B = ruv^T > 0$ and a corresponding positive eigenvector chosen is $u > 0$. Thus, $\lambda = 1 - rv^Tu \geq 0$ is the simple real eigenvalue of A of smallest modulo with eigenvector $u > 0$.

To compute the inverse A^{-1} under the condition $rv^Tu < 1$, we use the Sherman-Morrison formula, see Proposition 1.51. We find

$$A^{-1} = I + \frac{r}{1 - rv^Tu}\, uv^T \geq 0. \tag{2.5}$$

For $v = u \geq 0$ with $u^Tu > 0$, the matrices of the form (2.2), putting $r = (u^Tu)^{-1}r'$, lead to

$$A = I - r'\frac{uu^T}{u^Tu}. \tag{2.6}$$

Putting $r' = 2$, we derive the orthogonal Householder transformation matrix, see [18].

Lemma 2.11. now implies for A in (2.6):

    A is a nonsingular M-matrix for $0 \leq r' < 1$, i.e. A is Stieltjes matrix,

    A is a singular M-matrix for $r' = 1$,

    A is not an M-matrix for $r' > 1$.

2.2.2. Rank-two perturbations of the unit matrix

Our next aim is to consider special types of rank-two perturbations of the unit matrix which cover the bordered unit matrix.

Let $u \geq 0$, $v \geq 0$ be $(n-1)$-dimensional vectors. We form the nonnegative matrix B of order n as follows

$$B = \begin{pmatrix} 0 & u \\ v^T & 0 \end{pmatrix} \geq 0. \tag{2.7}$$

We assume B of rank-two and examine matrices A defined by

$$A = I - B. \tag{2.8}$$

In the next lemma, we impose restrictive conditions on u and v such that A in (2.8) is an M-matrix.

<u>Lemma 2.12.</u>

The matrix $A = I - B$, where $u \geq 0$ and $v \geq 0$, is an M-matrix if and only if

$$v^Tu \leq 1. \tag{2.9}$$

Proof. The spectrum of B is given by
$$\sigma(B) = \left\{ 0, +(v^Tu)^{1/2}, -(v^Tu)^{1/2} \right\}.$$

Thus, $S(B) = (v^T u)^{1/2}$ and $S(B) \leq 1 = s$ if and only if $v^T u \leq 1$.

$\square$

We remark that A is a nonsingular M-matrix if $v^T u < 1$ and it is a singular M-matrix if $v^T u = 1$.

Furthermore, we find

$$(I - B)^{-1} = \begin{pmatrix} I & -u \\ -v^T & 1 \end{pmatrix}^{-1} = \begin{pmatrix} I + \dfrac{uv^T}{1 - v^T u} & \dfrac{u}{1 - v^T u} \\ \dfrac{v^T}{1 - v^T u} & \dfrac{1}{1 - v^T u} \end{pmatrix} \geq 0.$$

The matrix $A = I - B$ is irreducible if and only if the vectors $u = (u_i)$ and $v = (v_i)$ do not contain any zero components, that is
$$\prod_{i=1}^{n} u_i v_i \neq 0.$$

Supposing that $u > 0$ and $v > 0$, then A is irreducible and $S(B) = (v^T u)^{1/2} > 0$. The corresponding eigenvector, chosen positive, is $u' = (u^T, (v^T u)^{1/2})^T > 0$, i.e. $Bu' = S(B)u'$ and $Au' = (1 - S(B))u'$.
Putting $v = u \geq 0$, the bordered unit matrix

$$A = \begin{pmatrix} I & -u \\ -u^T & 1 \end{pmatrix},$$

is a Stieltjes matrix if and only if $u^T u < 1$.

2.2.3. Triangular M-matrices

The simplest of all are perhaps the triangular M-matrices. That is, we consider matrices of the type $A = I - B$ of order n, where the matrix $B = (b_{ij}) \geq 0$ is a strictly lower ($b_{ij} = 0$ for $i \leq j$) or a strictly upper ($b_{ij} = 0$ for $i \geq j$) triangular matrix.
By this assumption, we have in any case $S(B) = 0$. Hence $A = I - B$ is a nonsingular M-matrix by $s = 1 > S(B) = 0$.
By Proposition 1.25. and, due to the fact that $B^k = 0$ for all integers $k \geq n$, we have

$$A^{-1} = (I - B)^{-1} = \sum_{k=0}^{n-1} B^k \geq 0.$$

Generalizing, we assume that $A = (a_{ij})$ is an upper or lower triangular matrix with $a_{ii} \geq 0$, $\forall i \in N$ and $a_{ij} \leq 0$, $i \neq j$. With a number $s \geq \max_{i \in N} a_{ii}$, we express A in the following form

$$A = sI - (sI - A) = sI - B,$$

where $B = sI - A \geq 0$. The eigenvalues of B are the diagonal entries of B such that $S(B) = \max_{i \in N} (s - a_{ii}) = s - \min_{i \in N} a_{ii} \geq 0$.

We conclude

$\min\limits_{i \in N} a_{ii} > 0$ implies $s > S(B)$, i.e. A is a nonsingular M-matrix,

$\min\limits_{i \in N} a_{ii} = 0$ implies $s = S(B)$, i.e. A is a singular M-matrix.

2.2.4. Tridiagonal M-matrices

Let $A = (a_{ij})$ be tridiagonal, i.e. $a_{ij} = 0$ for $|i - j| \geq 2$. For brevity, we use the following notation

$$A = \begin{pmatrix} c_1 & -b_1 & & & \\ -a_2 & c_2 & -b_2 & & \\ & \ddots & \ddots & \ddots & \\ & & & -a_n & c_n \end{pmatrix}, \tag{2.10}$$

and assume $c_i \geq 0$, $\forall i \in N$. Further, let $a_{i+1} > 0$, $b_i > 0$ for $i = 1,..,n-1$. The latter conditions are equivalent to the irreducibility of A. The reducible case, i.e. $\prod\limits_{i=1}^{n-1} b_i = 0$ or/and $\prod\limits_{i=2}^{n} a_i = 0$, can be reduced to a sequence of irreducible tridiagonal submatrices.

According to [19], there exists a positive diagonal matrix $D = \text{diag}(d_1,\ldots,d_n)$ such that $A' = D^{-1}AD$ is symmetric. The matrix D can be chosen such that

$$d_1 = 1, \quad d_i = \left(\frac{a_2 \cdots a_i}{b_1 \cdots b_{i-1}}\right)^{1/2}, \quad i = 2,\ldots,n.$$

From this it follows that

$$A' = D^{-1}AD = \begin{pmatrix} c_1 & -\sqrt{b_1 a_2} & & & \\ -\sqrt{b_1 a_2} & c_2 & -\sqrt{b_2 a_3} & & \\ & \ddots & \ddots & \ddots & \\ & & & -\sqrt{b_{n-1} a_n} & c_n \end{pmatrix} = (A')^T. \tag{2.11}$$

Because of $\sigma(A) = \sigma(A')$, the spectrum of A is real, and additionally, all of the $\lambda \in \sigma(A)$ are simple, see [19].

Lemma 2.13.

An irreducible tridiagonal matrix $A \in Z^{n \times n}$ is an M-matrix if and only if $\min\limits_{\lambda \in \sigma(A)} \lambda \geq 0$.

Proof. First, we remark that if $A \in Z^{n \times n}$ with diagonal entries $c_i \geq 0$, $\forall i \in N$ then, for the positive diagonal matrix D defined above, we have

$$A' = D^{-1}AD \in Z^{n \times n},$$

where the diagonal entries c_i remain unchanged.

24

Suppose for the proof

$$s \geq \max\left\{\max_{\lambda \in \mathsf{G}(A')} \lambda \; , \; \max_{i \in N} c_i\right\} ,$$

and express A' in the form $A' = sI - B$, where $B = sI - A' \geq 0$.
This implies $S(B) = \max\limits_{\lambda \in \mathsf{G}(A')} |s - \lambda| = \max\limits_{\lambda \in \mathsf{G}(A')} (s - \lambda) = s - \min\limits_{\lambda \in \mathsf{G}(A')} \lambda$.
Thus, $s \geq S(B)$ if and only if $\min\limits_{\lambda \in \mathsf{G}(A')} \lambda \geq 0$. Because of $\mathsf{G}(A) = \mathsf{G}(A')$,
the assertion is proven.

$\square$

The assertion of Lemma 2.13. leads us to the conclusion that a tridiagonal matrix $A \in Z^{n \times n}$ is a nonsinglar M-matrix if and only if
$\min\limits_{\lambda \in \mathsf{G}(A)} \lambda > 0$ and it is a singular M-matrix if $\min\limits_{\lambda \in \mathsf{G}(A)} \lambda = 0$.

In some special cases of tridiagonal matrices $A \in Z^{n \times n}$ we know the
whole spectrum $\mathsf{G}(A)$. This is the case if A is a tridiagonal
Toeplitz matrix, i.e.

$$A = \mathrm{tridiag}(-a,c,-b) \quad = \begin{pmatrix} c & -b & & \\ -a & c & -b & \\ & \ddots & \ddots & \ddots \\ & & -a & c \end{pmatrix}_{n \times n}, \qquad (2.12)$$

where $a \geq 0$ and $b \geq 0$. According to $[19,28]$, the spectrum $\mathsf{G}(A)$ is
given by

$$\mathsf{G}(A) = \left\{\lambda_k = c - 2\sqrt{ab}\, \cos\frac{k\pi}{n+1} \; , \; k \in N\right\}. \qquad (2.13)$$

From Lemma 2.13. it follows that A is a nonsingular M-matrix if
$c - 2\sqrt{ab} > 0$.
For $b = a$ and $c - 2a > 0$, the matrix A is a Stieltjes matrix. Then
the corresponding system of eigenvectors is

$$x_k = (\sin\frac{k\pi}{n+1}, \; \sin 2\frac{k\pi}{n+1}, \; \dots \, , \sin n\frac{k\pi}{n+1})^T \; , \; k \in N. \quad (2.14)$$

The eigenvector with all components of one sign is obviously x_1.

2.2.5. Nonsingular M-matrices which leave invariant the relations
        between the components of vectors

Mention should be made of those nonsingular M-matrices A which leave
invariant the relations between the components of any vector $x \in R^n$,
see $[49]$. This verbal statement is expressible in the following relations. Let A be a matrix which possesses the property mentioned
and let $y = (y_1,\dots,y_n)^T = Ax$ for $\forall x \in R^n$. Then

$$x_i - x_j = c\,(y_i - y_j) \qquad (2.15)$$

for $i \neq j$, where c is a positive constant.

The matrices which leave invariant the relations between the components of vectors are special types of nonsingular M-matrices. We now show this by constructing such matrices A.

Choosing n+1 real numbers $v_1 \geq 0, \ldots, v_n \geq 0, c > 0$, we define the positive number

$$d = c + \sum_{i=1}^{n} v_i > 0.$$

Next, with e being the vector of all ones and with $v = (v_1,\ldots,v_n)^T \geq 0$, we generate the following matrix A

$$A = \frac{1}{c}(I - \frac{1}{d} ev^T). \tag{2.16}$$

Up to the factor $1/c$, the matrix A is a rank-one perturbation of the unit matrix I, see Section 2.2.1. By Lemma 2.11., the matrix $I - \frac{1}{d}ev^T$ is a nonsingular M-matrix because of

$$1 - S(\frac{1}{d}ev^T) = 1 - \frac{1}{d} v^T e = 1 - \sum_{i \in N} v_i / (c + \sum_{i \in N} v_i) > 0.$$

Thus, A is a nonsingular M-matrix.

We now make use of the Sherman-Morrison formula, see Proposition 1.51., to compute the inverse A^{-1}. We have

$$A^{-1} = c I + e v^T \geq 0. \tag{2.17}$$

We show next that A from (2.16) leaves invariant the relations between the components of vectors. For any $x \in R^n$ let $y = Ax$. The conclusion is that

$$x = A^{-1}y = cy + ev^T y = cy + \sum_{k \in N} v_k y_k\, e.$$

From this it follows that

$$x_i = cy_i + \sum_{k \in N} v_k y_k ,$$

$$x_j = cy_j + \sum_{k \in N} v_k y_k .$$

Hence, on subtracting the two equations, we have the desired result

$$x_i - x_j = c (y_i - y_j).$$

In solving linear equation systems $Ax = y$, where A is of the form (2.16), the solution $x = A^{-1}y$ satisfies (2.15).

2.3. M-matrix conditions

In this section we shall be concerned with necessary and sufficient
or only with sufficient conditions for $A \in Z^{n \times n}$ to be a nonsingular
M-matrix. The literature on M-matrix conditions is rather extensive.
Therefore, we limit ourselves to gleaning a number of pertinent
statements from the numerous publications. The technical details of
the proofs are beyond the scope of the book and must therefore be
omitted.

Theorem 2.14. ,[2]

A matrix $A \in Z^{n \times n}$ is a nonsingular M-matrix if and only if A^{-1} exists
and $A^{-1} \geqslant 0$.

It is seen that Theorem 2.14. is equivalent to Definition 2.3. It
is also said that nonsingular M-matrices are inverse-positive or
inverse-monotone.

Theorem 2.15. ,[45]

A matrix $A \in Z^{n \times n}$ is a nonsingular M-matrix if and only if each
principal submatrix of A is a nonsingular M-matrix.

This result leads to the interpretation that the M-matrix property
is penetrated throughout an M-matrix.

Theorem 2.16. ,[45]

A matrix $A \in Z^{n \times n}$ is a nonsingular M-matrix if and only if the real
parts of all eigenvalues of A are positive.

Matrices A with Re $\lambda > 0, \forall \lambda \in \sigma(A)$ are called positive-stable. Thus
any nonsingular M-matrix A is positive-stable.

Theorem 2.17. ,[30]

A matrix $A \in Z^{n \times n}$ is a nonsingular M-matrix if and only if there
exists a nonsingular M-matrix X such that $A = X^2$.

Theorem 2.18. ,[2]

A symmetric matrix $A \in Z^{n \times n}$ is a Stieltjes matrix if and only if
A is positive definite.

Theorem 2.19. ,[29]

Let $A \in Z^{n \times n}$ be an L-matrix, where D = diag A. Then A is a nonsingular M-matrix if and only if

$$S(I - D^{-1}A) < 1. \tag{2.18}$$

For further conditions equivalent to the statement "A is a nonsingular M-matrix", we refer to [2,45].

It seems also important to have sufficient conditions which guarantee that a bordered M-matrix keeps the M-matrix property.
For this purpose, let A be a nonsingular M-matrix of order n-1.
Further, let $u \geqq 0$, $v \geqq 0$ be (n-1)-dimensional vectors and let $a > 0$ be a constant. We form the following bordered matrix A' of order n

$$A' = \begin{pmatrix} A & -u \\ -v^T & a \end{pmatrix} , \tag{2.19}$$

which is, obviously, an L-matrix.
We state

Theorem 2.20.

The matrix A' is a nonsingular M-matrix if and only if

$$v^T A^{-1} u < a. \tag{2.20}$$

Proof. By the assumptions, we have

$$A' = A_1' A_2' = \begin{pmatrix} A & 0 \\ 0^T & a \end{pmatrix} \begin{pmatrix} I & -u' \\ -(v')^T & 1 \end{pmatrix},$$

where $u' = A^{-1} u \geqq 0$, $v' = v/a \geqq 0$.
The matrix A_1' is a nonsingular M-matrix because of

$$(A_1')^{-1} = \begin{pmatrix} A^{-1} & 0 \\ 0^T & 1/a \end{pmatrix} \geqq 0.$$

By Lemma 2.12., the matrix A_2' is a nonsingular M-matrix if and only if $(v')^T u' < 1$, which is equivalent to condition (2.20).
Thus $\det A' = \det A_1' \det A_2' \neq 0$ and

$$(A')^{-1} = \begin{pmatrix} I + \dfrac{A^{-1}uv^T}{a-v^T A^{-1}u} & \dfrac{aA^{-1}u}{a-v^T A^{-1}u} \\ \dfrac{v^T}{a-v^T A^{-1}u} & \dfrac{a}{a-v^T A^{-1}u} \end{pmatrix} \begin{pmatrix} A^{-1} & 0 \\ 0^T & 1/a \end{pmatrix} =$$

$$= \begin{pmatrix} A^{-1} + \dfrac{A^{-1}uv^T A^{-1}}{a-v^T A^{-1}u} & \dfrac{A^{-1}u}{a-v^T A^{-1}u} \\ \dfrac{v^T A^{-1}}{a-v^T A^{-1}u} & \dfrac{1}{a-v^T A^{-1}u} \end{pmatrix} \geqq 0.$$

This explicit representation of $(A')^{-1}$ completes the proof. $\square$

Let us briefly consider the symmetric case $A' = (A')^T$. For this purpose, we assume A to be a Stieltjes matrix and choose $v = u \geqq 0$.

As a conclusion of Theorem 2.20., $A' = \begin{pmatrix} A & -u \\ -u^T & a \end{pmatrix}$ is a Stieltjes matrix if and only if $u^T A^{-1} u < a$ holds. It follows that

$$\det A' = (a - u^T A^{-1} u) \det A,$$

and we have $\det A' > 0$ because of $\det A > 0$ and condition (2.20) for $v = u$.

For given $u \geq 0$, $\neq 0$, we derive a lower bound for possible $a > 0$. That is,

$$a > \frac{u^T A^{-1} u}{u^T u} u^T u \geq \min_{\lambda \in \mathfrak{S}(A)} \frac{1}{\lambda} \cdot u^T u = \frac{1}{\lambda_{max}} u^T u \ ,$$

where $\lambda_{max} = \max\limits_{\lambda \in \mathfrak{S}(A)} \lambda$.

The use of Definition 2.1. usually causes the problem of good estimates of $S(B)$ to show whether $s \geq S(B)$ or not. On the other hand, if we make use of Definition 2.3., then under the assumption that $A \in Z^{n \times n}$ is nonsingular, we have to predict that all of the entries of A^{-1} are nonnegative.

Therefore, we add some practicable sufficient M-matrix conditions which only take into account the entries of A.

<u>Theorem 2.21.</u> ,[29]

Let A be an L-matrix which is strongly row or column diagonally dominant, i.e. $Ae > 0$ or $e^T A > 0^T$.
Then A is a nonsingular M-matrix.

Proof. Let $Ae > 0$. Setting $D = \text{diag } A$, we have $\| I - D^{-1} A \|_\infty < 1$. Thus, $S(I - D^{-1} A) < 1$ and the assertion follows by Theorem 2.19. If $e^T A > 0^T$, we consider A^T instead of A and use $(A^{-1})^T = (A^T)^{-1}$. The proof is complete. $\qquad\qquad\square$

If a symmetric matrix $A = A^T$ satisfies the assumptions of Theorem 2.21., then A is a Stieltjes matrix.

In the following theorem "strongly diagonally dominant" is replaced by "irreducible and weakly diagonally dominant".

<u>Theorem 2.22.</u> ,[29]

Let A be an irreducible L-matrix which is weakly row or column diagonally dominant, i.e. $Ae \geq 0$, $\neq 0$ or $e^T A \geq 0^T$, $\neq 0^T$.
Then A is a nonsingular M-matrix.

Proof. Let A be an irreducible L-matrix and $Ae \geq 0$, $\neq 0$. From the assumptions, we have $D = \text{diag } A$ is a positive diagonal matrix, the matrix $I - D^{-1}A$ is irreducible, $\| I - D^{-1}A \|_\infty = 1$ and for at least one $i \in N$

$$\frac{1}{a_{ii}} \sum_{j \neq i} |a_{ij}| < 1$$

holds.

Thus by Proposition 1.19., we have $S(I - D^{-1}A) < 1$. Applying now Theorem 2.19., it follows that A is a nonsingular M-matrix. If $e^T A \geqslant 0^T$, $\neq 0^T$, it suffices again to consider A^T instead of A in the proof. This completes the proof of the **theorem**. $\square$

We conclude from Theorem 2.22. that a symmetric irreducible and weakly diagonally dominant L-matrix is a Stieltjes matrix.

The irreducible diagonally dominant case of M-matrices is known as the Minkowski theorem.

Theorem 2.23. , Minkowski theorem, [47]

Let $B \geqslant 0$ be irreducible and $A = sI - B$, where $s > 0$ is a constant. From $(sI - B)e \geqslant 0$, $\neq 0$ or $(sI - B)^T e \geqslant 0$, $\neq 0$, it follows that $S(B) < s$.

As shown in [47] by an example of a 2×2-matrix, the converse of the Minkowski theorem is generally not true.

3. M - M A T R I X   P R O P E R T I E S

We now proceed to consider some main properties of M-matrices. They are of general interest, and besides they bear some direct relation-ship to discretization methods as will be seen later on. Referring to the literature, we shall omit the proofs, which are far from being elementary.
For surveys, the reader is referred to [2,21,26,28,29,45].

3.1. General properties

Property 3.1.

In general, the sum and the product of two M-matrices is not an M-matrix.
This property can be shown by the following simple example. Setting $A_1 = \begin{pmatrix} 1 & 0 \\ -a & 1 \end{pmatrix}$ and $A_2 = A_1^T$, where both matrices are nonsingular M-matrices for any $a \geqslant 0$, we find that the sum $A_1 + A_2$ is an M-matrix

only for $0 \leqslant a \leqslant 2$ (the sum is nonsingular for $0 \leqslant a < 2$ and singular for $a = 2$). For $A =$ tridiag $(-a,1,0)$ of order $n = 3$, which is a nonsingular M-matrix for any $a \geqslant 0$, $A^2 \in Z^{3 \times 3}$ holds only for $a = 0$. Therefore, A^2 is an M-matrix only for $a = 0$.

By special assumptions, the sum and the product of two M-matrices may be an M-matrix. Let us do some examples.

Example 3.1.1.

Let A_1 and A_2 be M-matrices which are strongly row diagonally dominant, that is, $A_i e > 0$, $i=1,2$. Then the sum $A_1 + A_2$ is an M-matrix which is also strongly row diagonally dominant, because of $(A_1 + A_2)e = A_1 e + A_2 e > 0$, see Theorem 2.21.

Example 3.1.2.

Let A_1 and A_2 be M-matrices where $A_1 A_2 \in Z^{n \times n}$. Then the product $A_1 A_2$ is an M-matrix, see [2].

Example 3.1.3.

The class of nonsingular M-matrices is closed under positive diagonal multiplication. In other words, if A is a nonsingular M-matrix and D is a positive diagonal matrix, which is also an M-matrix, then AD and DA are nonsingular M-matrices. To show this, we can directly apply Definition 2.3.

We remark that the product of two nonsingular M-matrices is in any case a monotone matrix.

Example 3.1.4.

It should be noted that the 2×2-M-matrices are closed under matrix multiplication. To show this property, we consider the following 2×2-M-matrices

$$A_1 = \begin{pmatrix} a_i & -b_i \\ -c_i & d_i \end{pmatrix} , \quad i = 1,2 .$$

Then we have $A_1 A_2 \in Z^{2 \times 2}$ by

$$A_1 A_2 = \begin{pmatrix} a_1 a_2 + b_1 b_2 & -(a_1 b_2 + b_1 d_2) \\ -(c_1 a_2 + d_1 c_2) & c_1 c_2 + d_1 d_2 \end{pmatrix} .$$

The assertion follows from Example 3.1.2.

Property 3.2.

The class of M-matrices is closed under permutation cogredient operation.

That is, for any M-matrix A and each permutation matrix P the matrix

$A' = P^T A P$ is also an M-matrix.

To prove this property, let $A = sI - B$ be an M-matrix. For each permutation matrix P we have $A' = sI - B'$, where $B' = P^T B P \geqslant 0$. Further, $\sigma(B) = \sigma(B')$, which implies that $S(B) = S(B')$. Therefore, $s \geqslant S(B) = S(B')$ proves the property considered.

Property 3.3.

All of the principal minors of a nonsingular M-matrix $A = (a_{ij})$ are positive, see [2].

This statement coincides with Theorem 2.4. according to principal minors of order one. A further inference is that $\det A > 0$ for any nonsingular M-matrix A.

Furthermore, we thus obtain the sums S_k of all the $k \times k$ principal minors of A positive, i.e. $S_k > 0$, $\forall k \in N$. From this property, it follows that all of the coefficients of the characteristic polynomial $\det(\lambda I - A)$ of a nonsingular M-matrix are nonzero and of alternating sign. That is

$$\det(\lambda I - A) = \lambda^n - S_1 \lambda^{n-1} + S_2 \lambda^{n-2} + \ldots + (-1)^{n-1} S_{n-1} \lambda + (-1)^n S_n. \tag{3.1}$$

By Theorem 2.16., all of the roots of the polynomial (3.1) have positive real parts.

Property 3.4.

The nonsingular M-matrices are latent strongly row diagonally dominant.

That is, for any nonsingular M-matrix $A = (a_{ij})$ of order n there exists a positive diagonal matrix $D = \text{diag}(d_1, \ldots, d_n)$ such that AD is strongly row diagonally dominant, i.e.

$$a_{ii} d_i > \sum_{j \neq i} |a_{ij}| d_j , \quad \forall i \in N. \tag{3.2}$$

By Example 3.1.3. the product AD is also a nonsingular M-matrix. It should be noted that for a given nonsingular M-matrix A the positive diagonal matrix D with $ADe > 0$ is not unique.

To show the existence of at least one D, we choose $d = A^{-1}e$. By Theorem 2.4. we have $d = (d_1, \ldots, d_n)^T > 0$ and define $D = \text{diag}(d_1, \ldots, d_n)$, where $d = De$. Hence

$$ADe = Ad = AA^{-1}e = e > 0.$$

We see, $\lambda = 1 \in \sigma(AD)$. The corresponding eigenvector is e. Further, if there exists a vector $x > 0$ such that $Ax > 0$, then with $D = \text{diag}(x_1, \ldots, x_n)$ and by $x = De$ we have

$$ADe = Ax > 0.$$

If we now repeat the considerations for A^T instead of A, it follows that for any nonsingular M-matrix A there exists a positive diagonal matrix $D' = \text{diag}(d_1', \ldots, d_n')$ such that $D'A$ is strongly column diagonally dominant, i.e. $e^T D'A > 0^T$.

In summary, for any nonsingular M-matrix A there exist positive diagonal matrices D and D' such that the M-matrix $D'AD$ is strongly row and column diagonally dominant, see [2,45].

Property 3.5.

For any nonsingular M-matrix A there exists a vector $x > 0$ such that $Ax > 0$.

We note that $x > 0$ is not unique, that is, the direction of x. We give some examples of possible $x > 0$ based on the previous considerations. First, x can be chosen $x = A^{-1}e > 0$, see Property 3.4. Secondly, let $A = sI - B$ be an irreducible nonsingular M-matrix. Then $x > 0$ may be the eigenvector of B which corresponds to the simple Perron-Frobenius eigenvalue $S(B)$, i.e. $Bx = S(B)x$, see Theorem 2.6. In this case we have

$$Ax = (sI - B)x = (s - S(B))x > 0.$$

The case of $x \geqq 0$ with $Ax > 0$ is also discussed in [45]. Here, we do not deal with it.

From the knowledge of a vector $x > 0$ such that $Ax \geq e$, it follows that

$$\| A^{-1} \|_\infty \leq \| x \|_\infty . \tag{3.3}$$

This property may be used for some practicable estimates of the norm $\| A^{-1} \|_\infty$.

To prove the inequality (3.3), we state for any vector $y \in R^n \setminus \{0\}$

$$|y| = (|y_1|, \ldots, |y_n|)^T \leq \| y \|_\infty e \leq \| y \|_\infty Ax .$$

Multiplying this inequality by $A^{-1} \geqq 0$ from the left, we find

$$A^{-1}|y| \leq \| y \|_\infty x .$$

By $|A^{-1}y| \leq A^{-1}|y|$ it follows that $|A^{-1}y| \leq \| y \|_\infty x$.

Hence

$$\frac{\| A^{-1}y \|_\infty}{\| y \|_\infty} \leq \| x \|_\infty ,$$

which was to be proved.

Let A be a strongly row diagonally dominant M-matrix. Then on comparison with Proposition 1.3. we can choose $x = (r_1^{-1},..,r_n^{-1})^T > 0$, which fulfils $Ax \geq e$, and we have again

$$\| A^{-1} \|_\infty \leq \max_{i \in N} \frac{1}{r_i} \quad .$$

Property 3.6.

Any nonsingular M-matrix $A = (a_{ij})$ does not reverse the sign of nonzero vectors, that is, if $x \neq 0$ and $y = Ax$, then for some subscribt $i \in N$, $x_i y_i > 0$, see [2].

The proof is carried out by assuming $N_+(x) \neq \emptyset$. Further, let $D = \mathrm{diag}(d_1,...,d_n)$ be a positive diagonal matrix such that

$$A' = (a'_{ij}) = AD \quad \text{with} \quad A'e = ADe > 0,$$

see Property 3.4.

Consider now $y = {}^\backprime DD^{-1}x = A'x'$, where $x' = D^{-1}x$. Let $x'_i = \max_{j \in N_+(x')} x'_j$. By $N_+(x') = N_+(x) \neq \emptyset$ we have $x'_i > 0$.

The i-th equation of $y = Ax$ takes the form

$$y_i = a'_{ii}x'_i + \sum_{j \in N_-(x')} a'_{ij}x'_j + \sum_{j \in N_+(x')} a'_{ij}x'_j . \qquad (3.4)$$

Multiplying this equation by $x'_i > 0$, we have

$$x_i y_i / d_i = x'_i y_i =$$

$$= a'_{ii}(x'_i)^2 + x'_i \sum_{j \in N_-(x')} a'_{ij}x'_j + x'_i \sum_{j \in N_+(x')} a'_{ij}x'_j \geq$$

$$a'_{ii}(x'_i)^2 + x'_i \sum_{j \in N_+(x')} a'_{ij}x'_j \geq (x'_i)^2 (a'_{ii} - \sum_{j \in N_+(x')} |a'_{ij}|) > 0.$$

Thus, $x_i y_i > 0$. This proves the first part of our assertion.

Next, let $N_-(x) = N_-(x') \neq \emptyset$. Suppose now that $x'_i = \min_{j \in N_-(x')} x'_j$ with $x'_i < 0$.

Then from equation (3.4) we have

$$x_i y_i / d_i \geq a'_{ii}(x'_i)^2 + x'_i \sum_{j \in N_-(x')} a'_{ij}x'_j \geq$$

$$(x'_i)^2 (a'_{ii} - \sum_{j \in N_-(x')} |a'_{ij}|) > 0.$$

Property 3.7.

Let A be an irreducible nonsingular M-matrix. Then $A^{-1} > 0$, see [26].

For the proof we consider A in the form $A = sI - B$, $s > 0$, $B \geq 0$, $s > S(B)$. This assumption leads us to A^{-1} via Proposition 1.25. Hence

$$A^{-1} = \frac{1}{s}(I - B/s)^{-1} = \sum_{k=0}^{\infty} \frac{1}{s^{k+1}} B^k.$$

The irreducibility of A implies B irreducible. Let $B^k = (b_{ij}^{(k)})$ for integers $k \geq 1$. By Proposition 1.10., for any pair (i,j) there exists an integer $k \leq n$, n the order of A, such that $b_{ij}^{(k)} > 0$.
Thus

$$\sum_{k=0}^{n} \frac{1}{s^{k+1}} B^k > 0 \quad \text{implies} \quad A^{-1} > 0.$$

Let us also note that if A is a reducible nonsingular M-matrix then, by Proposition 1.7., A^k is reducible for all integers $k = \pm 1, \pm 2, \ldots$. Thus, A^{-1} contains a number of zeros.

Recently several authors have discovered the interesting fact that the elements of the sequence $\{A^{-k}\}_{k=1}^{\infty}$ all have the same zero pattern. That is, for each pair (i,j) the ij-entry in A^{-k}, namely $a_{ij}^{(k)}$, is zero for all integers $k \geq 1$ or is positive for all this k. For more details we refer to [48].

Property 3.8.

For any nonsingular M-matrix A, there exists a monotone matrix $C \geq A$ such that $A' = I - C^{-1}A$ is convergent, see [45].

For a first example of C we refer to Theorem 2.19. If we put $C = \text{diag } A$, which is a monotone matrix, we have $C \geq A$ and $S(I - C^{-1}A) < 1$. We postpone considerations of further examples to later sections.

Property 3.9.

Let A be an irreducible M-matrix. Then the matrix $-A$ is essentially positive.

To see this, let $r \geq \max_{i \in N} a_{ii} \geq 0$. Then we have $rI - A \geq 0$ is irreducible and it is also primitive, by Proposition 1.31. Now it follows from Proposition 1.45. that $-A$ is an essentially positive matrix.

Thus,

$$0 \leq \exp(rI - A) = \exp(r) \exp(-A)$$

implies $\exp(-A) \geq 0$. The stronger inequality, $\exp(-tA) > 0$ for all $t > 0$, follows from Proposition 1.46.

3.2. Additive M-matrix perturbations

In the present section we discuss the additive perturbation problem
of M-matrices. That is, for an M-matrix A we consider the perturbed
matrix $A' = A + C$. The question is under which conditions on C the
matrix A' remains an M-matrix.

First, we investigate nonnegative perturbations of M-matrices, i.e.,
we assume $C \geq 0$. The main result is given in the following theorem.

<u>Theorem 3.10.</u> ,[21]

Suppose that $A = sI - B$ is an M-matrix. Then $A' = A + C$ is an
M-matrix for any $C = B - B' \geq 0$, where $0 \leq B' \leq B$.

Proof. By Proposition 1.18. it follows that $S(B') \leq S(B)$. Then, we
get $A' = A + C = sI - B'$, where $s > 0$, $B' \geq 0$ and $s \geq S(B) \geq S(B')$.
Hence, A' is an M-matrix. $\qquad\qquad\square$

Let us look at the nonnegative additive perturbation of nonsingular
M-matrices. We establish the following result.

<u>Theorem 3.11.</u> ,[21]

Let $A = sI - B$ be a nonsingular M-matrix. Under the conditions
$C = B - B' \geq 0$ with $0 \leq B' \leq B$, the matrix $A' = A + C$ is a nonsingular M-matrix and
$$(A')^{-1} \leq A^{-1}. \tag{3.5}$$

Proof. From the assumptions $s > S(B)$, $S(B) \geq S(B')$, we get $s > S(B')$.
Hence, $A' = A + C = sI - B'$ is a nonsingular M-matrix.
Further, by $B'/s \leq B/s$ we deduce that

$$(A')^{-1} = \frac{1}{s}(I - B'/s)^{-1} = \frac{1}{s}\sum_{k=0}^{\infty}\left(\frac{B'}{s}\right)^k \leq$$

$$\frac{1}{s}\sum_{k=0}^{\infty}\left(\frac{B}{s}\right)^k = \frac{1}{s}(I - B/s)^{-1} = A^{-1}.$$

The theorem is proved. $\qquad\qquad\square$

A special type of nonnegative perturbations of M-matrices is the
nonnegative diagonal perturbation. We state

<u>Theorem 3.12.</u>

Suppose that $A = sI - B$ is a nonsingular M-matrix. Then, for any
nonnegative diagonal matrix $C = \mathrm{diag}(c_1,\ldots,c_n) \geq 0$, the matrix
$A' = A + C$ is a nonsingular M-matrix such that
$$(A')^{-1} = (A + C)^{-1} \leq A^{-1}. \tag{3.6}$$

Proof. It is assumed here that $s > S(B)$. The matrix A' can be written
as $A' = sI - B + C = s'I - B'$, putting $s' = s + \max\limits_{i \in N} c_i$ and
$B' = \max\limits_{i \in N} c_i I - C + B$.

We have $s' \geq s > 0$ and $B' \geq B \geq 0$. Furthermore, $B' \leq \max\limits_{i \in N} c_i I + B$ holds.

By the monotonicity property of the spectral radius, see Proposition
1.18., we get

$$S(B') \leq S(\max\limits_{i \in N} c_i I + B) = \max\limits_{i \in N} c_i + S(B).$$

Thus, $s > S(B)$ implies $s + \max\limits_{i \in N} c_i > S(B) + \max\limits_{i \in N} c_i$, hence $s' > S(B')$,

which was to be proved.

To prove inequality (3.6), let $A = A' - C$, which is a regular splitting of the monotone matrix A. Thus, $S((A')^{-1}C) < 1$ by Proposition
1.50. It thus follows that

$$A^{-1} = (I - (A')^{-1}C)^{-1}(A')^{-1} = \left(\sum_{k=0}^{\infty} ((A')^{-1}C)^k \right)(A')^{-1}$$

$$= (A')^{-1} + \left(\sum_{k=1}^{\infty} ((A')^{-1}C)^k \right)(A')^{-1} \geq (A')^{-1},$$

because of $(A')^{-1}C \geq 0$. The theorem is proved. $\qquad \square$

It is readily verified that if A is an M-matrix, then, for any
$C = \text{diag}(c_1,...,c_n) \geq 0$, the matrix $A + C$ is also an M-matrix.

Suppose $C = B - B' \geq 0$ as in Theorem 3.10. and 3.11. or $C = \text{diag}(c_1,...$
$...,c_n) \geq 0$ as in Theorem 3.12., then the inequalities (3.5) and (3.6)
imply

$$\| (A + C)^{-1} \|_\infty \leq \| A^{-1} \|_\infty . \qquad\qquad (3.7)$$

The proof is evident.

As can be seen later on in the book, another type different from
$C \geq 0$ of additive M-matrix perturbations is of interest in connection
with discretization methods. We shall call it shortly L-perturbation
of M-matrices, because perturbation matrices C may also be L-matrices, see Definition 2.9.
To make this more precise, let

$$\mathscr{L}^{n \times n} = \left\{ C = (c_{ij}): c_{ii} \geq 0, \forall i \in N, c_{ij} \leq 0, i \neq j \right\}. \qquad (3.8)$$

It is obvious that $\mathscr{L}^{n \times n} \subset Z^{n \times n}$ and that any L-matrix is contained in $\mathscr{L}^{n \times n}$.

Conversely, there are $C \in \mathcal{L}^{n \times n}$ which are not L-matrices. For example, $C = \begin{pmatrix} 0 & -1 \\ 0 & 0 \end{pmatrix}$.

For our purpose, we define the following two subsets of $\mathcal{L}^{n \times n}$. That is

$$\mathcal{L}_r^{n \times n} = \left\{ C \in \mathcal{L}^{n \times n} : \quad Ce \geq 0 \right\}, \tag{3.9a}$$

$$\mathcal{L}_c^{n \times n} = \left\{ C \in \mathcal{L}^{n \times n} : \quad e^T C \geq 0^T \right\}. \tag{3.9b}$$

The subscript r stands for "row" and c indicates "column".

Then it is not very hard to show the following properties are valid. For any $A, B \in \mathcal{L}_r^{n \times n}$ and any real number $\alpha \geq 0$, we have $\alpha A \in \mathcal{L}_r^{n \times n}$ and $A + B \in \mathcal{L}_r^{n \times n}$.

We now prove the following results in respect of L-perturbations of M-matrices.

<u>Theorem 3.13.</u>

Suppose that A is a strongly row diagonally dominant M-matrix or that A is an irreducible weakly row diagonally dominant M-matrix. Then, for any $C \in \mathcal{L}_r^{n \times n}$, the matrix $A' = A + C$ is a nonsingular M-matrix.

Proof. Both of the assumptions imply, by Theorems 2.21. and 2.22., that A is a nonsingular M-matrix. Hence, A is also an L-matrix. First, let $Ae > 0$. Then, for any $C \in \mathcal{L}_r^{n \times n}$ it follows that $A'e = (A + C)e = Ae + Ce > 0$. Hence, A' is a strongly row diagonally dominant L-matrix. By Theorem 2.21., A' is then a nonsingular M-matrix.
Secondly, let A be irreducible with $Ae \geq 0, \neq 0$. Thus, A irreducible implies $A' = (a'_{ij}) = A + C$ irreducible for any $C \in \mathcal{L}_r^{n \times n}$, because $a'_{ij} \leq a_{ij} \leq 0$, $i \neq j$. On the other hand, we have $A'e = (A + C)e = Ae + Ce \geq 0, \neq 0$ such that A' is weakly row diagonally dominant. It is obvious that A' is an L-matrix. Hence, by Theorem 2.22., A' is a nonsingular M-matrix.
The theorem is proved. $\qquad\qquad\square$

We remark that Theorem 3.13. remains true for any $C \in \mathcal{L}_c^{n \times n}$, assuming the M-matrices A to be strongly column diagonally dominant or irreducible weakly column diagonally dominant. The proof remains the same as for Theorem 3.13., replacing A, C, A' by its transpose.

Furthermore, it is also relevant to note that for L-perturbations of nonsingular M-matrices an inequality similar to (3.5) or (3.6) does not generally hold. To illustrate this by an example, let

$$A = I, \quad C = \text{tridiag}(-1,1,0) \in \mathcal{L}_r^{n \times n}.$$

Then $A' = A + C = \text{tridiag}(-1,2,0) = (a'_{ij})$. Let $(A')^{-1} = ((a'_{ij})^-)$, we have $(a'_{ii})^- = \frac{1}{2}$ and

$$(a'_{ij})^- = \begin{cases} \text{positive} & \text{for } j < i \\ 0 & \text{for } j > i \end{cases}.$$

Thus $(A')^{-1} \leqslant A^{-1} = I$ is not true.

3.3. Factorization of M-matrices

At the beginning of this section a few remarks are in order. Here,
we do not discuss all the possible factorizations which are also
applicable to the factorization of M-matrices, see [28,29] and
elsewhere. In our considerations, we only call attention to M-matrix
factorizations which lead to products of M-matrices. Furthermore, we
assume that all the M-matrices under consideration are nonsingular.
Some aspects concerning the factorization of singular M-matrices
have been described, for instance, in [51].

To begin with, we state the result of the M-matrix factorization
into a product of triangular M-matrices, see [2,45,51].

<u>Theorem 3.14.</u>

For every nonsingular M-matrix A there exists the factorization into
the product of a lower and upper triangular M-matrix L and U, respectively. That is

$$A = LU, \tag{3.10}$$

where $L = (l_{ij})$, $l_{ii} > 0$, $\forall i \in N$, $l_{ij} = 0$, $i < j$, $l_{ij} \leqslant 0$, $i > j$,
and $U = (u_{ij})$, $u_{ii} > 0$, $\forall i \in N$, $u_{ij} \leqslant 0$, $i < j$, $u_{ij} = 0$, $i > j$.

For the proof of Theorem 3.14. we refer to [45].

The nonsingular triangular M-matrices L and U are L-matrices, see
Section 2.2.3.

It is then quite easy to see that the factorization (3.10) is not
unique. Let $D = \text{diag}(d_1,...,d_n) \geqslant 0$ be a positive diagonal matrix,
i.e. $\det D > 0$, we find

$$A = LU = LDD^{-1}U = L'U',$$

Where L' and U' are also lower and upper triangular nonsingular
M-matrices.

For a priori given $l_{ii} > 0$, $\forall\, i \in N$ or $u_{ii} > 0$, $\forall\, i \in N$ the factorization $A = LU$ of a nonsingular M-matrix A then is unique.
If A is a Stieltjes matrix, i.e. $A = A^T$ and positive definite, the LU-factorization becomes symmetric, that is $U = L^T$, such that

$$A = LL^T \tag{3.11}$$

is the unique Cholesky factorization, see [18].

Example 3.15.

As an example, consider the factorization of a tridiagonal nonsingular M-matrix A.
Let A be given by (2.10) in Section 2.2.4.

Setting $A_{-1} = 0$, $A_o = 1$ and defining the leading principal minors of A by

$$A_k = \det \begin{pmatrix} c_1 & -b_1 & & \\ -a_2 & c_2 & -b_2 & \\ & & \ddots & \\ & & -a_k & c_k \end{pmatrix}, \quad k=1,\dots,n ,$$

we find

$$A_k = c_k A_{k-1} - a_k b_{k-1} A_{k-2} , \quad k = 1,\dots,n . \tag{3.12}$$

By Property 3.3., it follows that $A_k > 0$, $k=1,\dots,n$.
Using (3.12) and putting $u_{ii} = 1$, $\forall\, i \in N$, we have

$$A = LU = \begin{pmatrix} \dfrac{A_1}{A_o} & & & \\ -a_2 & \dfrac{A_2}{A_1} & & \\ & -a_3 & \dfrac{A_3}{A_2} & \\ & & \ddots & \\ & & & -a_n & \dfrac{A_n}{A_{n-1}} \end{pmatrix} \begin{pmatrix} 1 & -b_1\dfrac{A_o}{A_1} & & & \\ & 1 & -b_2\dfrac{A_1}{A_2} & & \\ & & \ddots & \\ & & & 1 & -b_{n-1}\dfrac{A_{n-2}}{A_{n-1}} \\ & & & & 1 \end{pmatrix} . \tag{3.13}$$

For practical purposes, the LU-factorization (3.13) of a nonsingular M-matrix A is efficiently computable by the shortened Gaussian algorithm, see [23]. Thus, we find

$$A = LU = \begin{pmatrix} l_1 & & & \\ -a_2 & l_2 & & \\ & \ddots & \ddots & \\ & & -a_n & l_n \end{pmatrix} \begin{pmatrix} 1 & -\alpha_2 & & \\ & 1 & -\alpha_3 & \\ & & \ddots & \ddots \\ & & 1 & -\alpha_n \\ & & & 1 \end{pmatrix} ,$$

where

$$l_1 = c_1, \qquad \alpha_2 = b_1/l_1,$$
$$l_i = c_i - \alpha_i a_i, \qquad \alpha_{i+1} = b_i/l_i, \quad i=2,\ldots,n-1 \qquad (3.14)$$
$$l_n = c_n - \alpha_n a_n.$$

Assuming $Ae \geqslant 0$, $\neq 0$, then from [23] we know that the algorithm (3.14) is numerically stable.

Let us now turn to another type of factorization. It is based on the extraction of roots from M-matrices.

A detailed description of square roots of M-matrices may be found in [30]. Based on Definition 2.10., the following result holds.

<u>Theorem 3.16.</u> ,[30]

An M-matrix A has an M-matrix X as a square root, i.e. $A = X^2$, if and only if A has "property c".

From the comments on Definition 2.10. in Section 2.1., we conclude that any nonsingular M-matrix A has an M-matrix as a square root.

<u>Example 3.17.</u>

The assertion of Theorem 3.16. is the basis of the following factorization of a bordered nonsingular M-matrix A.

Suppose that A is a nonsingular M-matrix of order n-1 and let $u \geqslant 0$, $v \geqslant 0$ be (n-1)-dimensional vectors. Further, let $a > 0$ be a constant. Then, by Theorem 2.20., the bordered matrix

$$A' = \begin{pmatrix} A & -u \\ -v^T & a \end{pmatrix}$$

is a nonsingular M-matrix if and only if $v^T A^{-1} u < a$.

Under this assumption, we find the following factorization of A'

$$A' = \begin{pmatrix} A^{1/2} & 0 \\ -v^T A^{-1/2} & (a - v^T A^{-1} u)^{1/2} \end{pmatrix} \begin{pmatrix} A^{1/2} & -A^{-1/2} u \\ 0^T & (a - v^T A^{-1} u)^{1/2} \end{pmatrix} \qquad (3.15)$$

We remark that each of the factors in (3.15) is a nonsingular M-matrix. Furthermore, each of the factors is a rank-one perturbation of the nonsingular M-matrix

$$A = \begin{pmatrix} A^{1/2} & 0 \\ 0^T & a' \end{pmatrix} \quad \text{with} \quad a' = (a - v^T A^{-1} u)^{1/2} ,$$

provided that $u \neq 0$ and $v \neq 0$.

A much more general result is announced in [39]. That is, the class
of nonsingular M-matrices is closed under the extraction of arbitrary roots. This means, for any integer $k \geqslant 2$ and for each nonsingular M-matrix A, there exists an M-matrix X such that $A = X^k$. This
property was recognized only recently by several authors, see [39].

For a simple illustrative example let $A = \begin{pmatrix} 1 & 0 \\ -a & 1 \end{pmatrix}$, which is an nonsingular M-matrix for any $a \geqslant 0$. Then,

$$X = \begin{pmatrix} 1 & 0 \\ -a/k & 1 \end{pmatrix},$$

is the k-th root of A for each $k \geqslant 2$.

A third type of factorization of nonsingular M-matrices has been
described in the literature. The assumptions are weakened in the
sense that one of the factors is allowed to be only a monotone
matrix. We state the following result.

<u>Theorem 3.18.</u> , [45]

Suppose that A is a nonsingular M-matrix. Then there exists a monotone matrix A_1 and a nonsingular M-matrix A_2 such that $A = A_1 A_2$.

3.4. Maximum principles

Maximum principles should obviously be a property of mathematical
problems reasonably modelling physical, biological, economic or
technological processes. But the concept of "maximum principle" is
not consistently used in both the Theory of Differential Equations
and in Numerical Analysis. For examples we refer to [3,5,11,13,22,
23,24,27].
Discrete maximum principles are of importance in the study of approximations to differential equations, see [11,13,23,49,50,53].
In this section, we try to handle maximum principles as M-matrix
properties which are independent of the original problems to be
discretized. Nonsingular M-matrices obey some sort of maximum principles where in each case its row diagonally dominance property is
of significance.

3.4.1. Boundary maximum principle

The first sort of maximum principles, for brevity and definiteness
briefly called boundary maximum principle, establishes some very
close connections between differential equation problems with
Dirichlet's boundary conditions and its approximation by linear
equation systems involving nonsingular M-matrices. Several applications will be discussed in more detail later on in Chapter 4.

The maximum principle under consideration assumes a special structure of M-matrices.

Let A be an L-matrix of order $n > 2$ in the following manner

$$A = \begin{pmatrix} I & 0 \\ A_{21} & A_{22} \end{pmatrix}. \qquad (3.16)$$

Suppose that I is the unit matrix of order n' with $2 \leqslant n' < n$ and A_{22} is square of order $n'' = n - n'$. If A_{22} is a nonsingular M-matrix, then A is a nonsingular M-matrix by

$$A^{-1} = \begin{pmatrix} I & 0 \\ -A_{22}^{-1}A_{12} & A_{22}^{-1} \end{pmatrix} \geqslant 0 . \qquad (3.17)$$

Lemma 3.19.

Let A defined by (3.16) be an L-matrix. Further, let A_{22} be irreducible and weakly row or column diagonally dominant. Then A is a nonsingular M-matrix.

Proof. Using Theorem 2.22. and (3.17), the proof is trivial. $\square$

We remark that Lemma 3.19. holds if A_{22} is strongly row or column diagonally dominant.

Before formulating the boundary maximum principle, we describe the type of linear equation systems under consideration. Let $y = (y',y'')^T$, where $y' = (y_1,...,y_{n'})^T$ and $y'' = (y_{n'+1},...,y_n)^T$. By analogy, let $f = (f',f'')^T$, where it is assumed that $f'' = 0''$. Then we consider linear equation systems of the form

$$A y = \begin{pmatrix} I & 0 \\ A_{21} & A_{22} \end{pmatrix} \begin{pmatrix} y' \\ y'' \end{pmatrix} = \begin{pmatrix} f' \\ 0'' \end{pmatrix} = f . \qquad (3.18)$$

The boundary maximum principle for second order elliptic boundary value problems, see [22] , motivates the following definition.

Definition 3.20.

We say that the equation systems (3.18) with nonsingular M-matrix A satisfies the boundary maximum principle, if it implies, for its solution $y = (y_1,...,y_n)^T$, the inequalities

$$\min_{1 \leqslant k \leqslant n'} f_k \leqslant y_i \leqslant \max_{1 \leqslant k \leqslant n'} f_k, \quad \forall i \in N. \qquad (3.19)$$

Denote by e'' the vector of all ones of order $n'' = n - n'$. With this notation we state now the main theorem.

<u>**Theorem 3.21.**</u>

Suppose that the matrix (3.16) is a nonsingular M-matrix for which

$$A_{22} \quad \text{is irreducible,}$$

$$A_{21}^T \, e'' \; < \; 0', \tag{3.20}$$

$$(A_{21} \; A_{22}) \, e \; = \; 0'', \tag{3.21}$$

where $0 = (0', 0'')^T$. Then the equation system (3.18) satisfies the boundary maximum principle.

Proof. Let $y = (y_1, \dots, y_n)^T = A^{-1} f$ be the unique solution of (3.18) under the assumptions for A. Then, $y_i = f_i$ for $i = 1, \dots, n'$.

Further, condition (3.21) implies

$$y_i = \sum_{k \neq i} \left(- \frac{a_{ik}}{a_{ii}} \right) y_k = \sum_{\{k \in N:\, a_{ik} < 0\}} \left(- \frac{a_{ik}}{a_{ii}} \right) y_k \;, \quad i = n'+1, \dots, n. \tag{3.22}$$

By the irreducibility of A_{22}, we have $\{k \in N:\, a_{ik} < 0\} \neq \emptyset$ for all $i = n'+1, \dots, n$. Thus, each y_i, $i = n'+1, \dots, n$ is a convex linear combination of those solution components y_k for which $a_{ik} < 0$ holds. This is seen by

$$- \frac{a_{ik}}{a_{ii}} \; \geq \; 0, \quad \forall \, k \in N \setminus \{i\} \;,$$

and

$$\sum_{k \neq i} \left(- \frac{a_{ik}}{a_{ii}} \right) = \sum_{\{k \in N:\, a_{ik} < 0\}} \left(- \frac{a_{ik}}{a_{ii}} \right) = \; 1 \;.$$

The convex linear combination property now implies

$$\min_{\{k \in N:\, a_{ik} < 0\}} y_k \; \leq \; y_i \; \leq \; \max_{\{k \in N:\, a_{ik} < 0\}} y_k \;, \quad i = n'+1, \dots, n \;. \tag{3.23}$$

Let $f_{i_0} = \min_{1 \leq k \leq n'} f_k$. By assumption (3.20) there exists an i_0^*, $n'+1 \leq i_0^* \leq n$, for which $i_0 \in \{k \in N:\, a_{i_0^* k} < 0\}$. By Proposition 1.12., the associated directed graph $\mathcal{G}(A_{22})^0$ has a directed path which leads from vertex P_i to vertex $P_{i_0^*}$. If we use the inequalities

$$\min_{\{k \in N:\, a_{jk} < 0\}} y_k \; \leq \; y_j \;,$$

for each P_j on the path, we estimate y_i below up to $i_0 \in \{k \in N:\, a_{i_0^* k} < 0\}$. Thus, for any y_i, $i = n'+1, \dots, n$ there holds

$$\min_{1 \leq k \leq n'} f_k \; \leq \; y_i \;.$$

The proof of the upper bound of (3.19) is quite analogous. $\qquad \square$

Let $f \geqslant 0$ in the equation system (3.18). Then under the assumptions
of Theorem 3.21. for A, we have, for the solution $y = (y_1, \ldots, y_n)^T$
of (3.18)

$$0 \leqslant y_i \leqslant \max_{1 \leqslant k \leqslant n'} f_k , \quad \forall i \in N. \tag{3.24}$$

This latter estimate remains true if we change the assumptions of
Theorem 3.21. as follows.

Theorem 3.22.

Let the matrix (3.16) be a nonsingular M-matrix where it is assumed
that

$$A_{22} \text{ is irreducible,}$$
$$A_{21}^T e'' < 0',$$
$$(A_{21} \, A_{22}) \, e > 0''. \tag{3.25}$$

Further, let $f \geqslant 0$ in (3.18). Then the solution $y = (y_1, \ldots, y_n)^T$
of the equation system (3.18) satisfies (3.24).

Proof. First, we remark that $f \geqslant 0$ implies $y = A^{-1} f \geqslant 0$. Hence, we
confine our attention to the upper bound of (3.24).
For $i = 1, \ldots, n'$, we have $y_i = f_i \geqslant 0$. From (3.25), we get

$$a_{ii} > \sum_{k \neq i} (- a_{ik}) , \quad i = n'+1, \ldots, n.$$

Suppose that $a_{ii} = \varkappa_i \, \bar{a}_{ii}$, where $\varkappa_i > 1$ and

$$\bar{a}_{ii} = \sum_{k \neq i} (- a_{ik}) , \quad i = n'+1, \ldots, n.$$

Thus

$$y_i = \frac{1}{a_{ii}} \sum_{\{k \in N:\ a_{ik} < 0\}} (- a_{ik}) \, y_k = \frac{1}{\varkappa_i} \sum_{\{k \in N:\ a_{ik} < 0\}} \left(- \frac{a_{ik}}{\bar{a}_{ii}}\right) y_k \leqslant$$

$$\sum_{\{k \in N:\ a_{ik} < 0\}} \left(- \frac{a_{ik}}{\bar{a}_{ii}}\right) y_k \leqslant \max_{\{k \in N:\ a_{ik} < 0\}} y_k .$$

This estimate is based on the convex linear combination property,
because we have

$$- \frac{a_{ik}}{\bar{a}_{ii}} \geqslant 0, \quad \forall k \in N \setminus \{i\} \quad \text{and} \quad \sum_{k \neq i} \left(- \frac{a_{ik}}{\bar{a}_{ii}}\right) = 1 .$$

By the same argument used in the proof of Theorem 3.21., we conclude
that $y_i \leqslant \max_{1 \leqslant k \leqslant n'} f_k$, $i = n'+1, \ldots, n$ which proves the theorem. $\quad\square$

The above results may be summarized as follows. The condition $A_{21}^T \, e'' < 0'$ in the Theorems 3.21. and 3.22. is necessary for the access of the components of y'' to that of $y' = f'$.

In Theorem 3.22. the assumption (3.25) may be weakened such that $(A_{21} \; A_{22}) \, e \geq 0, \neq 0$. In the case of $a_{ii} = \sum_{k \neq i} (- a_{ik})$ we simply have $\varkappa_i = 1$, which is not out of the range of the proof of Theorem 3.22.

3.4.2. Region maximum principle

In the present section we introduce another type of maximum principles, briefly named region maximum principle, for equation systems involving strongly row diagonally dominant M-matrices, see [53]. On comparison with the boundary maximum principle, it also implies lower and upper solution bounds, for which sufficient conditions are deduced.

The region maximum principle is shown to be meaningful in connection with certain discretization methods for differential equation problems as will be seen in Chapter 4.

Suppose that we are given an equation system

$$A \, y \; = \; f, \tag{3.26}$$

where $A = (a_{ij})$ is a nonsingular M-matrix. Furthermore, motivated by several applications in Chapter 4, we assume that the right-hand side vector f is represented by

$$f \; = \; B \, v \; + \; \tau \, C \, w, \tag{3.27}$$

where $B = (b_{ij})$ and $C \doteq (c_{ij})$ are square matrices, $v = (v_1, \ldots, v_n)^T$, $w = (w_1, \ldots, w_n)^T$ and $\tau > 0$ is a constant.

In general, the right-hand side representation (3.27) is not unique and it is not necessary for B and C to be nonsingular.

Next, we introduce the region maximum principle by the following definition, see [53].

<u>Definition 3.23.</u>

We say that the equation system

$$A \, y \; = \; B \, v \; + \; \tau \, C \, w, \tag{3.28}$$

with nonsingular M-matrix A satisfies the region maximum principle if it implies, for the solution $y = (y_1, \ldots, y_n)^T$, the estimates

$$\min_{k \in N} v_k \; + \; \tau \min_{k \in N} w_k \; \leq \; y_i \; \leq \; \max_{k \in N} v_k \; + \; \tau \max_{k \in N} w_k \; , \; \forall i \in N. \tag{3.29}$$

We can now prove the following theorem.

Theorem 3.24.

Let A be a strongly row diagonally dominant M-matrix. Suppose that
$B \geq 0$, $C \geq 0$, $\tau > 0$ where

$$A e \; = \; B e \; = \; C e \; = \; (r_1, \ldots, r_n)^T > 0. \qquad (3.30)$$

Then the equation system (3.28) satisfies the region maximum
principle.

Proof. Let $y = (y_1, \ldots, y_n)^T$ be the unique solution of the equation
system (3.28). Assuming $y_{i_0} = \min\limits_{i \in N} y_i$, we obtain from the i_0-th
equation of the system (3.28)

$$a_{i_0 i_0} y_{i_0} \; = \; \sum_{k \neq i_0} (- a_{i_0 k}) \, y_k \; + \; \sum_{k \in N} b_{i_0 k} v_k \; + \; \tau \sum_{k \in N} c_{i_0 k} w_k,$$

the estimate

$$a_{i_0 i_0} y_{i_0} \; \geq \; y_{i_0} \sum_{k \neq i_0} (- a_{i_0 k}) \; + \; \min\limits_{k \in N} v_k \sum_{k \in N} b_{i_0 k} \; + \; \tau \min\limits_{k \in N} w_k \sum_{k \in N} c_{i_0 k}.$$

Using the assumptions (3.30), we find

$$r_{i_0} y_{i_0} \; \geq \; r_{i_0} \left(\min\limits_{k \in N} v_k \; + \; \tau \min\limits_{k \in N} w_k \right).$$

So far we have proved that

$$\min\limits_{k \in N} v_k \; + \; \tau \min\limits_{k \in N} w_k \; \leq \; y_i \; , \qquad \forall i \in N.$$

The proof of the upper bound remains the same which proves the
theorem. $\qquad\qquad\square$

It should be observed that the desired property of $Ae = r > 0$ is
not shared a priori by every nonsingular M-matrix A. As an example,
consider irreducible weakly row diagonally dominant M-matrices, that
is, $Ae \geq 0$, $\neq 0$.
Since any nonsingular M-matrix A is, by Property 3.4., latent
strongly row diagonally dominant, we can state the following result.

Theorem 3.25.

Let A be a nonsingular M-matrix and $D = \text{diag}\,(d_1, \ldots, d_n)$ be a positive diagonal matrix such that $ADe = r = (r_1, \ldots, r_n)^T > 0$. Then the
solution $y = (y_1, \ldots, y_n)^T$ of the equation system $Ay = f = (f_1, \ldots, f_n)^T$
satisfies

$$d_L \min_{k \in N} \frac{f_k}{r_k} \; \leq \; y_i \; \leq \; d_U \max_{k \in N} \frac{f_k}{r_k} \; , \quad \forall i \in N , \qquad (3.31)$$

where

$$d_L = \begin{cases} \max_{j \in N} d_j & \text{for} & \min_{k \in N} f_k \leq 0 \\ \min_{j \in N} d_j & \text{for} & \min_{k \in N} f_k \geq 0 \end{cases} , \quad d_U = \begin{cases} \max_{j \in N} d_j & \text{for} & \max_{k \in N} f_k \geq 0 \\ \min_{j \in N} d_j & \text{for} & \max_{k \in N} f_k \leq 0 \end{cases} . \qquad (3.32)$$

Proof. The existence of positive diagonal matrices D such that
$A'e = ADe = r > 0$ follows from Property 3.4.
Let $y' = D^{-1}y$, we have

$$Ay = ADD^{-1}y = A'y' = f .$$

Further, we choose $B = \mathrm{diag}(r_1, \ldots, r_n)$, $v = B^{-1}f$ and $C = B$, $w = 0$,
$\tau > 0$, i.e., we now consider the equation system $A'y' = Bv$.
It follows that $A'e = Be = Ce = r > 0$. By Theorem 3.24., we have

$$\min_{k \in N} \frac{f_k}{r_k} \; \leq \; y'_i \; = \; \frac{y_i}{d_i} \; \leq \; \max_{k \in N} \frac{f_k}{r_k}, \quad \forall i \in N.$$

Hence

$$d_i \min_{k \in N} \frac{f_k}{r_k} \; \leq \; y_i \; \leq \; d_i \max_{k \in N} \frac{f_k}{r_k} \; , \quad \forall i \in N.$$

This, together with the definition of d_L and d_U in (3.32), completes
the proof of Theorem 3.25. $\qquad\qquad\qquad\qquad\qquad\qquad\square$

It is clear that Theorem 3.25. includes the case of strongly row
diagonally dominant M-matrices for which we can choose $D = I$.
The upper and lower bound of the enclosure (3.31) is strict, as the
following example shows.

Example 3.26.

Suppose $a_i > 0$, $\forall i \in N$ and consider

$$A = \begin{pmatrix} a_1 & & & & \\ -a_1 & a_2 & & \bigcirc & \\ & -a_2 & a_3 & & \\ & & \ddots & \ddots & \\ & & & -a_{n-1} & a_n \end{pmatrix} .$$

'Then, A is a nonsingular M-matrix and $Ae > 0$ fails if we do not
assume $0 < a_1 < a_2 < \ldots < a_n$.
Let

$$D = \mathrm{diag}(\frac{1}{a_1}, \frac{2}{a_2}, \ldots, \frac{n}{a_n}) ,$$

we find $ADe = e = r > 0$, that is, $A' = AD$ is then a strongly row diagonally dominant M-matrix.

The solution of $Ay = f = (f_1,...,f_n)^T$ is given by

$$y_i = \frac{1}{a_i} \sum_{j=1}^{i} f_j , \quad \forall i \in N.$$

Now, let us choose $f = e$, then the solution components are $y_i = \frac{1}{a_i}$, $\forall i \in N$. Thus, $f_k/r_k = 1$, $\forall k \in N$.
Then, by virtue of (3.31), we find

$$\min_{j \in N} \frac{j}{a_j} \leq y_i = \frac{1}{a_i} \leq \max_{j \in N} \frac{j}{a_j} , \quad \forall i \in N,$$

which is strict.

In connection with the preceding discussion, we derive a rule of finding positive diagonal matrices D for certain irreducible weakly row diagonally dominant M-matrices such that $A'e = ADe > 0$.
From the Properties 3.4. and 3.5. we can see the following. Any vector $x = (x_1,...,x_n)^T > 0$ such that $Ax > 0$ leads directly to a positiv diagonal matrix $D = \text{diag}(x_1,...,x_n)$ with $ADe > 0$. Conversely, any positive diagonal matrix $D = \text{diag}(d_1,...,d_n)$ with $ADe > 0$ immediately defines the vector $x = De > 0$ such that $Ax > 0$.

Our next aim is to realize the latent strongly row diagonally dominant property for certain nonsingular weakly row diagonally dominant M-matrices. In other words, we have to construct vectors $x > 0$ such that $Ax > 0$ and define then $D = \text{diag}(x_1,...,x_n)$.

Example 3.27.

Suppose that the symmetric Toeplitz matrix

$$A = \begin{pmatrix} a_o & -- & a_p & & & \\ & & & \diagdown & & \\ a_p & & & & & \\ & & & & & a_p \\ & & & & & \\ & & & a_p & -- & a_o \end{pmatrix}_{n \times n} \tag{3.33}$$

is a nonsingular M-matrix. This assumption implies that $a_o > 0$ and $a_k \leq 0$ for $k=1,...,p$. Further, let $n \geq 2p+1$ and

$$a_o = 2 \sum_{k=1}^{p} (- a_k) .$$

Hence, A is a weakly diagonally dominant M-matrix, i.e. $Ae \geq 0, \neq 0$.

In the following we need Jensen's inequality, see [8].

Let (a,b) be an arbitrary interval. Then, for any concave function $\varphi(x) \in C^2(a,b)$, that is $\varphi''(x) \leq 0$ for any $x \in (a,b)$, Jensen's inequality

$$\varphi\left(\frac{\sum s_k x_k}{\sum s_k}\right) \geq \frac{\sum s_k \varphi(x_k)}{\sum s_k} \tag{3.34}$$

holds, where $x_k \in (a,b)$, $s_k > 0$ and every sum in (3.34) ranges over a well defined index set $\{k\}$. In (3.34) the strong inequality holds for strongly concave functions $\varphi(x)$, i.e., if $\varphi''(x) < 0$ for each $x \in (a,b)$.

To apply Jensen's inequality, we set $h = \frac{1}{n+1}$ and $x_i = ih$, $i = -p, \ldots, n+p+1$. Putting $a = x_{-p}$ and $b = x_{n+p+1}$, let $\varphi(x) \in C^2(a,b)$ be a function with the following properties:

$$\varphi''(x) < 0 \quad \text{for each } x \in (a,b), \tag{3.35a}$$

$$\varphi(x) > 0 \quad \text{for each } x \in (a,b). \tag{3.35b}$$

Examples of such functions $\varphi(x)$ are easily available, for instance, we can choose $\varphi(x) = -x^2 + c$, c sufficiently large.

The desired vector x is then defined by

$$x = (\varphi(x_1), \ldots, \varphi(x_n))^T > 0, \tag{3.36}$$

which satisfies $Ax > 0$.

For the proof, we remark that for any $i \in N$ we have $x_{i+k} = x_i + kh$. By

$$1 = 2 \sum_{k=1}^{p} \left(-\frac{a_k}{a_o}\right),$$

we find

$$x_i = 2x_i \sum_{k=1}^{p}\left(-\frac{a_k}{a_o}\right) = \sum_{k=1}^{p}\left(-\frac{a_k}{a_o}\right) x_{i-k} + \sum_{k=1}^{p}\left(-\frac{a_k}{a_o}\right) x_{i+k} .$$

From (3.34) and (3.35a) we have

$$\varphi(x_i) = \varphi\left(\sum_{k=1}^{p}\left(-\frac{a_k}{a_o}\right) x_{i-k} + \sum_{k=1}^{p}\left(-\frac{a_k}{a_o}\right) x_{i+k}\right) >$$

$$\frac{1}{a_o}\left(\sum_{k=1}^{p}(-a_k)\varphi(x_{i-k}) + \sum_{k=1}^{p}(-a_k)\varphi(x_{i+k})\right).$$

Thus

$$\sum_{k=1}^{p} a_k \varphi(x_{i-k}) + a_o \varphi(x_i) + \sum_{k=1}^{p} a_k \varphi(x_{i+k}) > 0. \tag{3.37}$$

For i=p+1,...,n-p, (3.37) is the desired inequality. Deleting from
(3.37), for i=1,...,p and for i=n-p+1,...,n, all the nonpositive expressions $a_k \varphi(x_j)$, for which $j < 1$ and $j > n$, respectively, we get
the desired inequalities, which show that $Ax > 0$. $\qquad\qquad$ $\square$

If we change the assumptions of Theorem 3.24. in the following
manner, we can deduce only one-sided bounds for the solution of
the equation system (3.28).

<u>Theorem 3.28.</u> , [53]

Suppose that A is a strongly row diagonally dominant M-matrix, i.e.
$Ae = (r_1,...,r_n)^T > 0$. Let $B \geq 0$, $C \geq 0$ and $\tau > 0$.
Then, if $v \geq 0$, $w \geq 0$ and

$$\max_{i \in N} r_i \quad \leq \quad \min_{k \in N} \left\{ (Be)_k, (Ce)_k \right\} = c_{min}, \qquad (3.38)$$

the solution $y = (y_1,...,y_n)^T$ of the equation system (3.28) satisfies

$$\min_{k \in N} v_k + \tau \min_{k \in N} w_k \leq y_i , \quad \forall i \in N. \qquad (3.39)$$

Proof. From the assumptions it follows that $f = Bv + \tau Cw \geq 0$,
thus $y = A^{-1} f \geq 0$.
Let $y_{i_o} = \min_{i \in N} y_i$. Considering the i_o-th equation of the system
(3.28), we find

$$a_{i_o i_o} y_{i_o} \geq y_{i_o} \sum_{k \neq i_o} (- a_{i_o k}) + c_{min} \left(\min_{k \in N} v_k + \tau \min_{k \in N} w_k \right).$$

Hence

$$r_{i_o} y_{i_o} \geq c_{min} \left(\min_{k \in N} v_k + \tau \min_{k \in N} w_k \right) ,$$

so from $c_{min} \geq r_{i_o}$, we get (3.39). $\qquad\qquad$ $\square$

In analogy to Theorem 3.28. we state the next theorem.

<u>Theorem 3.29.</u> , [53]

Let A be an M-matrix as in Theorem 3.28. Further let $B \geq 0$, $C \geq 0$
and $\tau > 0$. Then, if $v \geq 0$, $w \geq 0$ and

$$\min_{i \in N} r_i \geq \max_{k \in N} \left\{ (Be)_k, (Ce)_k \right\} = c_{max} > 0, \qquad (3.40)$$

the solution y of the equation system (3.28) satisfies

$$y_i \leq \max_{k \in N} v_k + \tau \max_{k \in N} w_k , \quad \forall i \in N. \qquad (3.41)$$

Proof. From the assumptions, we have again $y = A^{-1}f \geq 0$.

Let $y_{i_o} = \max\limits_{i \in N} y_i$. From the i_o-th equation of (3.28), we find

$$a_{i_o i_o} y_{i_o} \leq y_{i_o} \sum_{k \neq i_o} (- a_{i_o k}) + c_{max} \left(\max\limits_{k \in N} v_k + \tau \max\limits_{k \in N} w_k \right).$$

Thus

$$r_{i_o} y_{i_o} \leq c_{max} \left(\max\limits_{k \in N} v_k + \tau \max\limits_{k \in N} w_k \right)$$

and the assertion follows from $0 < c_{max} \leq r_{i_o}$.

From the assumptions, we have $y \geq 0$, and $y_{i_o} = 0$ implies $y = 0$. For $y = 0$ the inequality (3.41) is true. $\qquad\square$

We now turn to nonsingular weakly row diagonally dominant M-matrices A and show, rather than compute a positive diagonal matrix D such that $ADe > 0$, how to apply the region maximum principle for deducing a lower solution bound of equation system (3.28).

We then have the following theorem.

<u>Theorem 3.30.</u> , [53]

Let A be a nonsingular weakly row diagonally dominant M-matrix, i.e. $Ae \geq 0, \neq 0$. Further, let $B \geq 0, C \geq 0$ with

$$r_i = (Be)_i = (Ce)_i > 0, \quad \forall i \in N \tag{3.42}$$

and let $v \geq 0$, $w \geq 0$ and $\tau > 0$. If there exists a diagonal matrix $D = \mathrm{diag}(d_1, \dots, d_n) \geq 0$ such that

$$r_i = (Ae)_i + d_i, \quad \forall i \in N, \tag{3.43}$$

then the equation system (3.28) implies for its solution $y = (y_1, \dots, y_n)^T$

$$\min\limits_{k \in N} v_k + \tau \min\limits_{k \in N} w_k \leq y_i, \quad \forall i \in N. \tag{3.44}$$

Proof. Instead of system (3.28) we consider the perturbed system

$$(A + D)z = Bv + \tau Cw = f, \tag{3.45}$$

where $f \geq 0$ by the assumptions made above and $A + D$ is a nonsingular M-matrix by Theorem 3.12. According to (3.42) and (3.43) the matrix $A + D$ is strongly row diagonally dominant by $(A + D)e = (r_1, \dots, r_n)^T > 0$. Applying now Theorem 3.24. to equation system (3.45), we find

$$\min\limits_{k \in N} v_k + \tau \min\limits_{k \in N} w_k \leq z_i \leq \max\limits_{k \in N} v_k + \tau \max\limits_{k \in N} w_k, \quad \forall i \in N. \tag{3.46}$$

Finally, from (3.6), we have $z = (A + D)^{-1}f \leq A^{-1}f = y$ which completes the proof of the theorem. $\qquad\square$

3.4.3. Maximum principle for inverse column entries

Up to now we have considered two maximum principles for equation
systems involving nonsingular M-matrices based on different enclosures of the solution components.
In the present section we introduce another sort of maximum principles which reflect important properties of the inverse $A^{-1} \geqslant 0$ of
nonsingular M-matrices A. In connection with discretization methods,
we derive in Chapter 4 examples of M-matrices A which satisfy the
maximum principle for inverse column entries and where its inverse
$A^{-1} \geqslant 0$ **takes** the shape of Green's functions of the discretized
differential operators, see also Chapter 6.

According to [49,50] , we give the following definition.

<u>Definition 3.31.</u>

We say a matrix A satisfies the maximum principle for inverse
column entries if

$$A\,y \quad = \quad f, \tag{3.47}$$

with $f \geqslant 0,\ \neq 0$, imply $y \geqslant 0$ and, moreover

$$\max_{i \in N} y_i \quad = \quad \max_{i \in N_+(f)} y_i. \tag{3.48}$$

In general, Definition 3.31. concerns monotone matrices A. But by
Theorem 2.2., every nonsingular M-matrix A is a monotone matrix.
Thus, Definition 3.31. is also applicable to nonsingular M-matrices.

The interpretation of (3.48) shows that if A satisfies the maximum
principle for inverse column entries, then the maximum response
takes place in such a part of the solution y of Ay = f where there
is a nonzero influence of the right-hand side vector f.

Now we explain our interpretation of the maximum principle under
consideration. For this, let $f = e_i$, the i-th coordinate unit vector,
in (3.47). Then, $y = A^{-1} e_i$ is exactly the i-th column of the inverse
$A^{-1} = (a_{ij}^-) \geq 0$. Hence, the inverse A^{-1} is, by Definition 1.4.,
weakly diagonally dominant of its column entries. That is

$$a_{ii}^- \geq a_{ji}^- , \quad \forall\, j \in N, \tag{3.49}$$

and for every $i \in N$.

The reader wishing to acquire a deeper understanding of the maximum
principle given by Definition 3.31., is referred to [50].

We shall confine our consideration to the case of nonsingular
M-matrices A. The following result is proved.

Theorem 3.32. , [50]

Let A be a nonsingular M-matrix. Then $Ae \geqslant 0, \neq 0$ is necessary and
sufficient for the maximum principle for inverse column entries.

Next, several examples of nonsingular M-matrices are given to illustrate the introduced maximum principle.

Example 3.33.

Let $A = \text{tridiag}(-a,1,0)$ be of order n. From Section 2.2.3. we have,
the triangular matrix A is a nonsingular M-matrix for each $a \geqslant 0$.
It is then quite easy to see that A^{-1} is the Toeplitz matrix given
by

$$
A^{-1} = \begin{pmatrix}
1 & & & & \\
a & 1 & & & \\
a^2 & a & 1 & & \\
\vdots & & & \ddots & \\
a^{n-1} & \cdots & a^2 & a & 1
\end{pmatrix} \geq 0. \tag{3.50}
$$

By Theorem 3.32., the matrix A satisfies the maximum principle for
inverse column entries if and only if $0 \leqslant a \leqslant 1$. For $0 \leqslant a < 1$, the
inverse A^{-1} is strongly diagonally dominant of its column entries,
but for $a = 1$ it is only weakly diagonally dominant of its column
entries.

Example 3.34.

Let $A = \text{tridiag}(-1,2,-1)$ be of order n. It then follows by Theorem
2.22. that A is a nonsingular M-matrix. Furthermore, we have $Ae \geqslant 0, \neq 0$.
Now, from Theorem 3.32., we conclude that A satisfies the maximum
principle for inverse column entries. Let $A^{-1} = (a^-_{ij})$, the entries
of A^{-1} are given by the following formula, see [26]

$$
a^-_{ij} = \begin{cases}
\dfrac{n - i + 1}{n + 1} \, j , & j < i , \\[2ex]
\dfrac{n - j + 1}{n + 1} \, i , & j \geqslant i .
\end{cases} \tag{3.51}
$$

Hence, $A^{-1} > 0$ is strongly diagonally dominant if its column entries.
Additionally we remark that $\det A = n+1$ for each $n \geqslant 1$.

Example 3.35.

Let

$$
A = \begin{pmatrix}
1 & -1 & & & \\
-1 & 2 & -1 & & \\
& & \ddots & & \\
& & -1 & 2 & -1 \\
& & & -1 & 2
\end{pmatrix}_{n \times n} .
$$

Then A is a nonsingular M-matrix by Theorem 2.22. The conditions $Ae \geqslant 0, \neq 0$ of Theorem 3.32. are fulfilled, so A satisfies the maximum principle for inverse column entries. Let $A^{-1} = (a^-_{ij})$, then

$$a^-_{ij} = \begin{cases} n - j + 1 \, , & j < i \, , \\ n - i + 1 \, , & j \geqslant i \, , \end{cases} \qquad (3.52)$$

see [26].

Hence, $A^{-1} > 0$ is weakly diagonally dominant of its column entries. Furthermore, it may be noted that $\det A = 1$ for each $n \geqslant 1$.

In [47] the following result is given.

<u>Theorem 3.36.</u> , Metzler's theorem

If A is an irreducible nonsingular M-matrix with $Ae \geqslant 0$ then for its inverse $A^{-1} = (a^-_{ij})$ holds

$$0 < a^-_{ji} \leqslant a^-_{ii} \, , \qquad \forall \, i,j \in N \, . \qquad (3.53)$$

From the Examples 3.34. ans 3.35. we observe that the inverse $A^{-1} = (a^-_{ij})$ of an irreducible nonsingular tridiagonal M-matrix A which satisfies the maximum principle for inverse column entries does not only share (3.49) but does even much more than this.

We derive the following result.

<u>Theorem 3.37.</u>

Let

$$A = \begin{pmatrix} c_1 & -b_1 & & & \\ -a_2 & c_2 & -b_2 & & \\ & \ddots & \ddots & \ddots & \\ & & & -a_n & c_n \end{pmatrix}$$

be an irreducible nonsingular M-matrix and assume that $Ae \geqslant 0, \neq 0$. Then, for its inverse $A^{-1} = (a^-_{ij})$, we have

$$a^-_{ii} \geqslant a^-_{i+1i} \geqslant \ldots \geqslant a^-_{ni} > 0 \, , \quad i=1,\ldots,n-1, \qquad (3.54a)$$

and

$$0 < a^-_{1i} \leqslant a^-_{2i} \leqslant \ldots \leqslant a^-_{ii}, \quad i=2,\ldots,n. \qquad (3.54b)$$

Proof. The irreducibility of A implies $\prod\limits_{i=1}^{n-1} b_i > 0$, $\prod\limits_{i=2}^{n} a_i > 0$ and $A^{-1} > 0$, see Property 3.7.

Further, by Theorem 3.32., we have $a^-_{ii} \geqslant a^-_{ji}$, for each $i,j \in N$.

Using now the assumptions

$$c_1 \geqslant b_1 > 0 ,$$
$$c_i \geqslant a_i + b_i > 0, \quad i = 2,..,n-1 ,$$
$$c_n \geqslant a_n > 0 ,$$

we shall prove (3.54a).

It follows from $AA^{-1} = I$ that

$$- a_n a^-_{n-1\,i} + c_n a^-_{ni} = 0 \quad \text{for } i=1,..,n-1.$$

Thus,

$$a^-_{n-1\,i} = a^-_{ni} c_n / a_n \geqslant a^-_{ni} .$$

Suppose now that

$$0 < a^-_{ni} \leqslant a^-_{n-1\,i} \leqslant \ldots \leqslant a^-_{ji} \quad \text{for some } j > i.$$

Then, we have

$$- a_j a^-_{j-1\,i} + c_j a^-_{ji} - b_j a^-_{j+1\,i} = 0 ,$$

hence

$$a^-_{j-1\,i} = (c_j a^-_{ji} - b_j a^-_{j+1\,i})/a_j \geqslant a^-_{ji} (c_j - b_j)/a_j \geqslant a^-_{ji} .$$

Thus, the proof of (3.54a) is complete. The proof of (3.54b) remains the same. This completes the proof. $\qquad\square$

Consider now the case where a nonsingular M-matrix A satisfies the maximum principle for inverse column entries and where A^{-1} is additionally strongly diagonally dominant of its column entries.

From [47] we have the following theorem.

Theorem 3.38.

Let $A = sI - B$ be a nonsingular M-matrix, where $B \geqslant 0$ and $Ae > 0$. Then, for $A^{-1} = (a^-_{ij})$, we have

$$0 < a^-_{ji} < a^-_{ii} , \quad \text{for } \forall j \neq i ,$$

and for each $i \in N$.

4. M-MATRICES AND DISCRETIZATION METHODS

There are two different views of looking at discretization methods
for the numerical solution of boundary and initial boundary value
problems. The first one focusses attention on the convergence analysis of the methods used, the second primarily investigates how
the applied methods reflect basic properties of continuous problems
in discrete approximations. For second order linear elliptic and
parabolic problems, which we shall consider in the following, these
properties are, for instance, inverse monotonicity, nonnegativity
and monotonicity of solutions, maximum principles and conservation
laws. Of course, there exist close relationships between these two
approaches to discretization methods.

The methods considered here for the numerical solution of the above
mentioned types of problems are finite difference methods (FDM),
finite element methods (FEM) and methods of lines (ML). These are
powerful and highly successful numerical methods, which have been
used to obtain approximate solutions to a wide variety of problems
in mathematics and engineering. The literature of them is extensive.

Confining our attention to qualitative properties of FDM, FEM and
ML, we shall study an intimate relationship between nonsingular
M-matrices and the deduced discrete approximations. In order to
illustrate the main ideas, we shall discuss representative examples
of problems, omitting all the details of the application of the discretization methods.

4.1. Problems

At the outset, let us briefly characterize the types of continuous
problems under consideration.

Let $x \in R^d$, $d \geq 1$. Suppose that Ω is a bounded connected region, i.e.
$\Omega \subset R^d$, with $\partial\Omega$ (for $d > 1$), the boundary of Ω , being sufficiently smooth. Roughly speaking, the smoothness of $\partial\Omega$ is to guarantee
the partial integration over Ω , that is

$$\int_\Omega \frac{\partial u}{\partial x_i} \, v \, dx \; = \; - \int_\Omega u \, \frac{\partial v}{\partial x_i} \, dx \; + \; \int_{\partial\Omega} u \, v \, \cos(\nu, x_i) \, ds, \qquad (4.1)$$

for $i = 1,\ldots,d$, where ν denotes the outward normal to Ω and
(ν, x_i) is the angle between ν and the positive direction of the

x_1-axis. For example, if $\partial\Omega$ is Lipschitzian, then (4.1) holds for any $u,v \in W_2^1(\Omega)$, see [4].

To state the problems, we begin by defining the following linear differential expression

$$L u \;=\; -\nabla(\, k\,\nabla u + b\,u\,) + q\,u\,. \tag{4.2}$$

Suppose that the scalar functions $k(x)$, $q(x)$ and the vector function $b(x) = (b_1(x),..,b_d(x))^T$ for $d > 1$, otherwise a scalar function $b(x)$, are sufficiently smooth over the region Ω . Further, let $k(x) \geqslant k_o = \text{const} > 0$ and $q(x) \geqslant 0$ for each $x \in \Omega$. Thus, L together with corresponding boundary conditions on $\partial\Omega$ is a linear elliptic differential operator.

The book deals with the following three types of problems.

First, let $u = u(x)$ for $x \in \Omega$.

The linear elliptic boundary value problems we consider are

$$
\begin{aligned}
L u &= f\,, & \Omega,\\[4pt]
\alpha u + \beta \frac{\partial u}{\partial \nu} &= g\,, & \partial\Omega.
\end{aligned}
\tag{4.3}
$$

Further, we consider linear eigenvalue problems

$$
\begin{aligned}
L u &= \lambda u\,, & \Omega,\\[4pt]
u &= 0\,, & \partial\Omega.
\end{aligned}
\tag{4.4}
$$

Second, let $u = u(x,t)$ for $(x,t) \in \Omega \times (0,T)$, the space-time cylinder over Ω , $0 < T < \infty$. Then, the linear parabolic initial boundary value problem under consideration is defined by

$$
\begin{aligned}
\frac{\partial u}{\partial t} + L u &= f\,, & \Omega \times (0,T]\,,\\[4pt]
u(x,0) &= u_o(x)\,, &\\[4pt]
\alpha u + \beta \frac{\partial u}{\partial \nu} &= g(x,t)\,, & \partial\Omega \times (0,T]\,.
\end{aligned}
\tag{4.5}
$$

In the sequel, we specify the more general problems (4.3) through (4.5) to apply FDM, FEM or ML, respectively, for its numerical solution.

According to problem (4.3) and (4.5) we assume that all its input data f,g,u_o,α,β guarantee a unique solution in the classical or generalized sense. The assumption on the eigenvalue problem (4.4) is that in this case L is a self-adjoint positive definite operator.

We turn now to a brief characterization of the FDM, FEM and ML under consideration.

Suppose that we are given an FDM grid $\overline{\omega}_h \subset \overline{\Omega}$ (or $\overline{\omega}_h \subset \overline{\Omega \times (0,T)}$) or an FEM subdivision of $\overline{\Omega}$ with corresponding set of nodal points $\overline{\omega}_h \subset \overline{\Omega}$. The subscript h indicates the discretization step size. We remark that $\overline{\omega}_h = \omega_h + \gamma_h$ where $\omega_h \subset \Omega$ and $\gamma_h \subset \partial\Omega$ with $\omega_h \cap \gamma_h = \emptyset$.

From the application of the FDM or FEM to the problems (4.3) or (4.5) we obtain a discrete operator L_h (or $(\frac{\partial}{\partial t} + L)_h$), an approximate solution u_h and a right-hand side f_h.

Thus, we have

$$\left.\begin{array}{c} (4.3) \\ (4.5) \end{array}\right\} \xrightarrow{\text{FDM, FEM}} L_h u_h = f_h . \qquad (4.6)$$

We identify the discrete operator L_h with a matrix A, i.e. $L_h \equiv A$, besides, the approximate solution u_h with a vector y, i.e. $u_h \equiv y$, and the right-hand side f_h with a vector f, $f_h \equiv f$. Under these assumptions, we can rewrite (4.6) as a linear equation system

$$A y = f . \qquad (4.7)$$

For brevity, the subscript h is omitted in (4.7). The order $n = n(h)$ of the system is well defined as will be seen later on in the examples.

According to the entries a_{ij} of A we shall assume the following sign pattern

$$a_{ii} > 0, \ \forall i \in N \quad \text{and} \quad a_{ij} \leq 0, \ i \neq j . \qquad (4.8)$$

Hence, A is at least an L-matrix, see Definition 2.9.

Moreover, we assume that A is a nonsingular M-matrix.

This assumption is quite natural in many cases. For example, let $\mathscr{S}(i)$ be the approximation star of the FDM for solving problems such as (4.3) numerically. The finite difference equations may arise in the following form

$$a_{ii} y_i + \sum_{j \in \mathscr{S}(i)} a_{ij} y_j = f_i , \ \forall x_i \in \omega_h, \qquad (4.9)$$

where

$$a_{ii} > 0 \quad \text{and} \quad a_{ij} < 0 \text{ for each } j \in \mathscr{S}(i),$$

and

$$a_{ii} + \sum_{j \in \mathscr{S}(i)} a_{ij} \geq 0 , \ \forall x_i \in \omega_h. \qquad (4.10)$$

Then the equations (4.9) complemented by the discretized boundary
conditions of (4.3) or (4.5) form the difference equation system
(4.7), where A is a weakly or strongly row diagonally dominant M-matrix.

The matrices A in (4.7) usually are sparse. That is, the number of
nonzero entries in each row and column of A is essentially less than
the order n of A.

We call the matrices A finite difference or finite element matrices.

Concerning the eigenvalue problem (4.4), we shall consider the finite
dimensional eigenvalue problem for A, where A by our assumptions is
a Stieltjes matrix, see Definition 2.8.

Thus, we have

$$(4.4) \xrightarrow{\text{FDM, FEM}} A y = \mu y . \qquad (4.11)$$

The application of the ML to the approximation of the initial boundary value problem (4.5) leads to a Cauchy problem for an ordinary
differential equation system. Let $y = y(t) = (y_1(t),\dots,y_n(t))^T$,
$0 \leqslant t \leqslant T$.

We get

$$(4.5) \xrightarrow{\text{ML}} \quad \dot{y} + A y = f,\ t > 0, \quad y(0) = y_o, \qquad (4.12)$$

where A is a nonsingular M-matrix.

In the following sections, we shall treat the numerical solution of
the three types of problems by FDM, FEM and ML from the point of
view how these methods carry over basic properties of the continuous
problems into the discrete approximations.

4.2. Irreducibility of discretized problems

The irreducibility of the matrices A, see (4.7), arising in the
application of discretization methods is of importance to the in-
vestigation of its inherent properties.

Suppose that the region $\bar{\Omega}$ is discretized by an FDM grid

$$\bar{\omega}_h = \left\{ x_i \in \bar{\Omega} , \quad i=1,\dots,n \right\} = \omega_h + \gamma_h, \qquad (4.13)$$

or that $\bar{\Omega}$ is subdivided into finite elements, where the set of
nodal points is also denoted by $\bar{\omega}_h$. Then, for every $x_i \in \omega_h$ there
is a corresponding approximation star $\mathscr{S}(i)$, see for instance (4.9).
Further, we shall assume that for each $x_i \in \omega_h$ we have to determine
one and only one approximate value y_i.

The problem of deciding whether or not the generated matrix $A = (a_{ij})$ is irreducible can be managed by building the associated directed graph $\mathcal{G}(A)$, see Definition 1.11. For this, all $x_i \in \overline{\omega}_h$ form the n vertices of the graph $\mathcal{G}(A)$. Then, all directed edges in $\mathcal{G}(A)$ are well-defined by the approximation star $\mathscr{S}(i)$, $x_i \in \omega_h$. That is, a directed edge leads from vertex x_i to vertex x_j, $i \neq j$, if and only if $a_{ij} \neq 0$. Hence, by virtue of Proposition 1.12., the matrix A is irreducible if and only if $\mathcal{G}(A)$ is strongly connected. In many cases, such a conclusion can be easily obtained.

We look briefly at some examples.

One-dimensional case.
Let $\Omega = (a,b)$ be an interval and let

$$\overline{\omega}_h = \left\{ x_i : \ a = x_1 < x_2 < \ \ldots \ < x_n = b \right\}.$$

Suppose that we obtain a discrete approximation, for instance of a problem of type (4.3), as a linear equation system $Ay = f$, where A is tridiagonal, that is

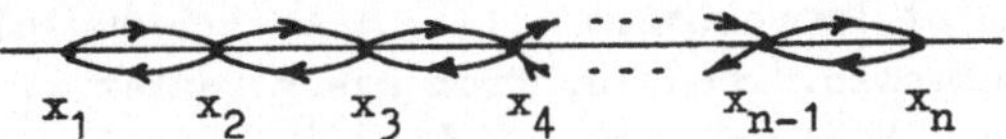

$$A \quad = \quad \begin{pmatrix} \ \end{pmatrix} .$$

Then A is irreducible if and only if the associated directed graph $\mathcal{G}(A)$ is such as shown in Figure 4.1.

Fig.4.1.

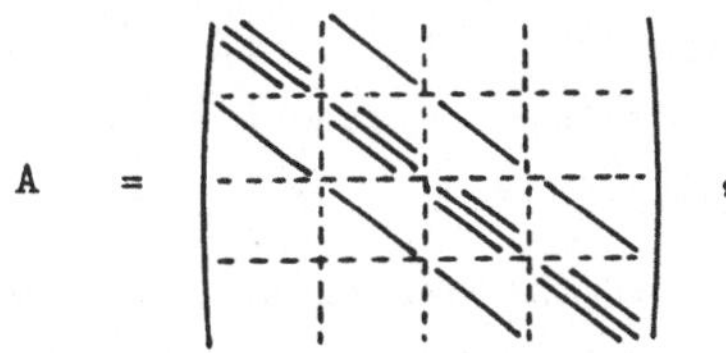

Two-dimensional case.
Let $\Omega = (a_1,b_1) \times (a_2,b_2)$ be a rectangular region and let

$$\overline{\omega}_h = \left\{ x_{ij} = (ih_1, jh_2), \ i = 0,..,n_1, \ j = 0,..,n_2, \ h_k = (b_k - a_k)/n_k, \ k = 1,2 \right\}.$$

Suppose that the resulting equation system $Ay = f$ has the following block tridiagonal structure

$$A \quad = \quad \begin{pmatrix} \ \end{pmatrix} ,$$

where all entries of A indicated by a line are nonzero. This is, for instance, the case if we apply FDM to solve numerically the Laplace equation $- \Delta u = 0, \ \Omega$ with Dirichlet boundary conditions $u = g, \partial\Omega$

and assume a corresponding numbering of the unknowns y_{ij}. The associated directed graph $\mathcal{G}(A)$ is shown in Figure 4.2.

Fig.4.2.

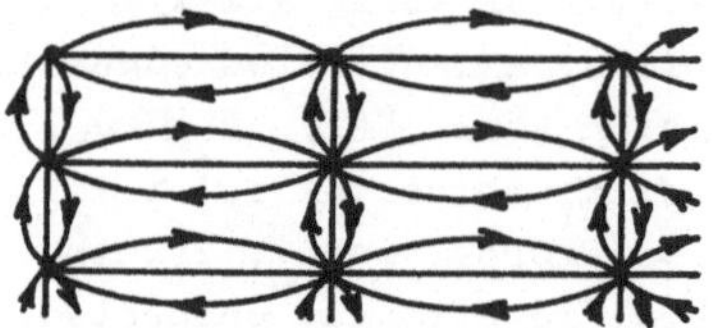

Lastly, let Ω be a general polygonal region and let $\overline{\omega}_h$ be defined by a triangulation of $\overline{\Omega}$. Then the associated directed graph $\mathcal{G}(A)$ may be such as shown in Figure 4.3.

Fig.4.3.

We remark that in nonsymmetric discrete problems, i.e. $A \neq A^T$, the structure of the nonzero entries of A often remains symmetric. That is, $a_{ij} \neq 0$ if and only if $a_{ji} \neq 0$, $i \neq j$. Then, a directed edge $\overrightarrow{x_i x_j}$ in $\mathcal{G}(A)$ implies the directed edge $\overrightarrow{x_j x_i}$ and conversely. Thus, it suffices to connect the vertices x_i and x_j by an undirected edge $\overline{x_i x_j}$. Hence, the associated graph $\mathcal{G}(A)$ is no longer directed. Therefore, by virtue of Proposition 1.12., A is irreducible if and only if $\mathcal{G}(A)$ is connected. That is, from every vertex x_i to every other vertex x_j there exists a path in $\mathcal{G}(A)$ which connects x_i and x_j. Thus $\mathcal{G}(A)$ can be identified with the grid $\overline{\omega}_h$ and its connections of grid points according to the used approximation stars $\mathcal{S}(i)$, $\forall x_i \in \omega_h$.

For a given discrete approximation which yields the equation system $Ay = f$, any renumbering of the unknowns can be expressed by $\tilde{y} = Py$ for a certain permutation matrix P.

Then, let A be a nonsingular M-matrix. If we consider, instead of $Ay = f$, the equivalent transformed system $\tilde{A}\,\tilde{y} = \tilde{f}$, where

$$\tilde{A} = P A P^T, \qquad \tilde{f} = P f,$$

then the M-matrix property of the system matrix $\tilde{A}$ remains unchanged by Property 3.2.

4.3. Finite difference methods

In the present section we show how nonsingular finite difference
matrices qualifying as M-matrices may reflect essential properties
of continuous problems in FDM approximations. Further we demonstrate by examples that FDM schemes cease to be stable and oscilla-
tions occur in the approximate solutions if the M-matrix property
of the finite difference matrices fails.
It will always be assumed that the problems under consideration are
unique solvable and its solution is sufficiently smooth.

4.3.1. Three-point difference approximations to one-dimensional elliptic boundary value problems

Suppose for definiteness $\Omega = (0,1)$ with $\partial\Omega = \{0,1\}$. Let $\overline{\Omega}$ be
discretized by the grid

$$\overline{\omega}_h = \{x_i: \ 0 = x_0 < x_1 < \ldots < x_{n-1} < x_n = 1\} = \omega_h + \gamma_h,$$

where

$$\gamma_h = \{x_0, x_n\}, \qquad h_i = x_i - x_{i-1}, \quad \forall i \in N.$$

We define the three-point difference star by $\mathscr{S}(i) = \{x_{i-1}, x_i, x_{i+1}\}$,
$\forall x_i \in \omega_h$. Any grid function y over $\overline{\omega}_h$ will immediately be interpreted as vector $y = (y_0, \ldots, y_n)^T$.
For sake of simplicity, we usually use uniform grids $\overline{\omega}_h$, letting
$h_i = h = 1/n$, $\forall i \in N$, i.e. $x_i = ih$.

First of all, we consider the following model problem.

<u>Problem 4.1.</u>

$$Lu = -u'' + b(x) u' = 0, \quad x \in \Omega,$$

$$u(0) = u_0, \quad u(1) = u_1.$$

Suppose that $b(x)$ is a bounded function in Ω. Then any C^2 solution
$u(x) \neq const$ satisfies the boundary maximum principle

$$\min \{u_0, u_1\} \leq u(x) \leq \max \{u_0, u_1\}, \quad \forall x \in \overline{\Omega}, \qquad (4.14)$$

see [22]. If there exists $\int b(x)dx$, then it follows from [16]
that

$$u(x) = C_1 + C_2 \int \exp(\int b(x)) \, dx, \qquad (4.15)$$

where the constants C_i, $i=1,2$ are determined by the Dirichlet
boundary conditions of Problem 4.1.

From (4.15) we deduce that u(x) is a monotone function, that is,
u(x) is monotone decreasing or increasing in dependence of $u_o > u_1$
or $u_o < u_1$, respectively.

In particular, let $b(x) = b = const \neq 0$. Then Problem 4.1. has the
solution

$$u(x) = ((u_1 - u_o)\exp(bx) + u_o\exp(b) - u_1)/(\exp(b) - 1). \qquad (4.16)$$

Putting $\varepsilon = 1/b$ in the case $|b| \gg 1$, we get the following singu-
larly perturbed problem.

Problem 4.1.'

$$Lu = -\varepsilon u'' + u' = 0, \quad x \in \Omega ,$$

$$u(0) = u_o , \quad u(1) = u_1 .$$

The solution of Problem 4.1.' involves a boundary layer near x = 0
if $\varepsilon < 0$ or near x = 1 if $\varepsilon > 0$, see [7]. The behaviour of u(x)
for $u_o = 0$ and $u_1 = 1$ is shown in Figure 4.4.

Fig.4.4.

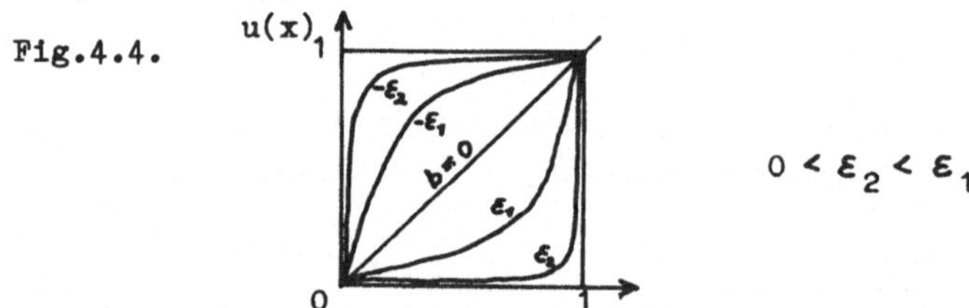

There is extensive literature on FDM discretizations for the model
Problems 4.1. and 4.1.'. From this field we shall pick out typical
situations for presenting our point of view.

Let us first consider the following FDM approximation to **Problem 4.1.**

Approximation 4.1.a   (method of centered differences)

The finite difference equation system, on a uniform grid $\overline{\omega}_h$, is
defined by

$$y_o = u_o ,$$

$$-D_+D_-y_i + b_iD_oy_i = 0 , \quad i=1,\dots,n-1 ,$$

$$y_n = u_1 ,$$

where $b_i = b(x_i)$.

This approximation is of second order accuracy, see [23].

Let $\gamma_i = hb_i/2$ for i=1,..,n-1. Then we build the matrix form $Ay = f$,
where

$$A = \begin{pmatrix} 1 & 0 & & & \\ -(1+\gamma_1) & 2 & -(1-\gamma_1) & & \\ & -(1+\gamma_2) & 2 & -(1-\gamma_2) & \\ & & \ddots & \ddots & \ddots \\ & & -(1+\gamma_{n-1}) & 2 & -(1-\gamma_{n-1}) \\ & & & 0 & 1 \end{pmatrix}, \qquad (4.17)$$

$$f = (u_0,0,\dots,0,u_1)^T = u_0 e_1 + u_1 e_{n+1} .$$

Given the permutation matrix

$$P = \begin{pmatrix} 1 & 0 & -\!-\!-\!-\!-\!- & 0 & 0 \\ 0 & 0 & & 0 & 1 \\ 0 & 1 & & 0 & 0 \\ \vdots & & \ddots & & \vdots \\ 0 & 0 & -\!-\!-\!-\!-\!- & 1 & 0 \end{pmatrix}_{(n+1)\times(n+1)} , \qquad (4.18)$$

we transform $Ay = f$ into $A'y' = f'$ where $A' = PAP^T$, $y' = Py$ and $f' = Pf$. Then A' is a matrix of the form (3.16), that is,

$$A' = \left(\begin{array}{cc|c} 1 & 0 & \\ 0 & 1 & \text{\Large O} \\ \hline A'_{21} & & A'_{22} \end{array} \right) . \qquad (4.19)$$

We observe that $Ae = P^T A'e \geqq 0, \neq 0$, which implies $A'e \geqq 0, \neq 0$. Furthermore, it is easily seen that A and A' are L-matrices if and only if

$$|\gamma_i| = h|b_i|/2 \leq 1 \quad \text{for } i=1,\dots,n-1. \qquad (4.20)$$

We remark that A is a nonsingular M-matrix under the assumptions (4.20).

From now we impose

$$|\gamma_i| < 1 \quad \text{for } i=1,\dots,n-1 , \qquad (4.21)$$

which holds for sufficiently small h. Then A'_{22} in (4.19) is an irreducible nonsingular M-matrix. It has been found that the assumptions of Theorem 3.21. hold and thus Approximation 4.1.a satisfies the discrete boundary maximum principle. That is, in analogy to (4.14), we have

$$\min \{u_0,u_1\} \leq y_i \leq \max \{u_0,u_1\} \quad \text{for } i=0,\dots,n . \qquad (4.22)$$

Furthermore, the approximate solution y is strongly monotone if $u_0 \neq u_1$. In the case $u_0 < u_1$, we have

$$u_0 = y_0 < y_1 < y_2 < \dots < y_{n-1} < y_n = u_1 , \qquad (4.23)$$

and

$$u_0 = y_0 > y_1 > y_2 > \dots > y_{n-1} > y_n = u_1 \qquad (4.24)$$

if $u_0 > u_1$.

To show this, we consider the i-th difference equation and rewrite
it as

$$y_i = \alpha_i y_{i-1} + \beta_i y_{i+1} , \qquad (4.25)$$

where

$$\alpha_i = (1+\gamma_i)/2 > 0, \quad \beta_i = (1-\gamma_i)/2 > 0, \quad \alpha_i + \beta_i = 1 . \qquad (4.26)$$

From this it follows that each y_i for $i=1,..,n-1$ is a convex
linear combination of its neighbours y_{i-1} and y_{i+1}. Hence

$$\min\{y_{i-1},y_{i+1}\} \leqslant y_i \leqslant \max\{y_{i-1},y_{i+1}\} , \qquad (4.27)$$

for $i=1,..,n-1$. This excludes the appearance of local minima
and maxima in y. Further, $y_i = y_{i+1}$ for some i would imply that
$y_{i-1} = y_i$ and $y_{i+1} = y_{i+2}$, which is impossible since $u_o \neq u_1$ was
assumed.
Thus only $u_1 = u_o$ implies $y = u_o e = $ const.

From the above considerations we can deduce much more detailed
information on the shape of the approximate solution y.

First, consider $b(x) \equiv 0$, $x \in (0,1)$. Thus $\gamma_i = 0$ for $i=1,..,n-1$
and the solution y of the difference equation system $Ay = f$ is
linear, that is

$$y_i = u_o + ih(u_1 - u_o) \quad \text{for } i=0,..,n .$$

This means that the approximation yields the exact solution at
all grid points $x_i \in \overline{\omega}_h$.

In analogy to convex and concave functions, we say that a vector
$y = (y_o,...,y_n)^T$ is strongly convex or strongly concave if

$$y_i < \frac{y_{i-1} + y_{i+1}}{2} \qquad \text{or} \qquad y_i > \frac{y_{i-1} + y_{i+1}}{2} ,$$

for $i=1,..,n-1$, respectively. We say that y is weakly convex or
weakly concave if " $\leqslant$ " or " $\geqslant$ " holds, respectively.

Suppose now for definiteness that $u_o < u_1$, hence we have $y_i < y_{i+1}$
for $i=0,..,n-1$. Let $b(x) > 0$, $x \in \Omega$. Then the solution y of $Ay = f$
is strongly convex. To show this, we have from (4.25) and (4.26)

$$y_i = \frac{y_{i-1} + y_{i+1}}{2} - \gamma_i(y_{i+1} - y_{i-1})/2 < \frac{y_{i-1} + y_{i+1}}{2} ,$$

for $i=1,..,n-1$. If $b(x) < 0$, $x \in \Omega$, y is strongly concave.
Both situations are depicted in Figure 4.5.

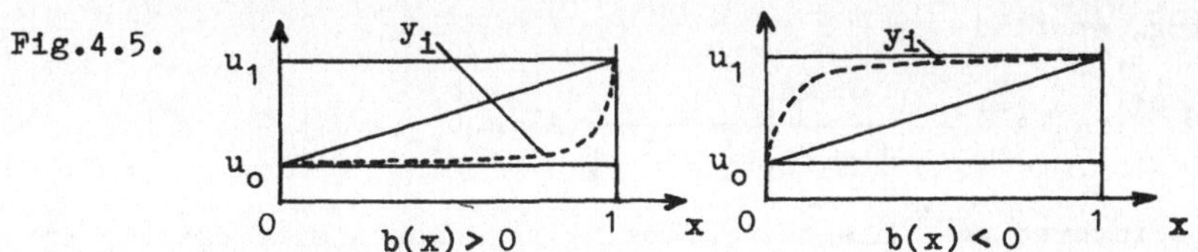

If $b(x)$ changes sign at some $x^* \in \Omega$, then the solution y of $Ay = f$ near x^* changes from convex to concave behaviour, or conversely.

The application of Approximation 4.1.a to the boundary layer Problem 4.1.´ immediately shows that all the above deduced properties for the approximate solution y to Problem 4.1.´ remain true under the assumption

$$| \gamma_i | \; = \; \frac{h}{2|\varepsilon|} \; < \; 1 \; . \tag{4.28}$$

The restrictions (4.21) and (4.28) are stability conditions for the centered difference method, which requires h sufficiently small. It is well-known that the method of centered differences on a uniform grid ceases to be stable and oscillations occur in y if $| \gamma_i | > 1$, see [7,37].

Let us now analyse such situations more carefully. For this suppose let

$$A \; = \; \begin{pmatrix} c_1 & -b_1 & & & \\ -a_2 & c_2 & -b_2 & & \\ & \ddots & \ddots & \ddots & \\ & & & -a_n & c_n \end{pmatrix} \tag{4.29}$$

be an irreducible strongly or weakly row diagonally dominant M-matrix. From the assumptions, we have

$$c_i > 0 \;\; \text{for each } i \in N, \qquad b_i > 0, \; a_{i+1} > 0 \;\; \text{for } i=1,..,n-1,$$
$$Ae \geq 0, \; \neq 0, \quad A^{-1} > 0.$$

Now we consider the following perturbations $\mathcal{A}$ of A and its inverse $\mathcal{A}^{-1}$ where $\mathcal{A}$ is no longer an M-matrix.

(a) Let $2 \leq k \leq n$ and

$$\mathcal{A} \; = \; A \; + \; 2a_k e_k e_{k-1}^T \; . \tag{4.30}$$

Then $\mathcal{A}$ has at position $(k,k-1)$ the entry $a_k > 0$ and all other entries of A are unchanged in $\mathcal{A}$. By the Sherman-Morrison formula, see Proposition 1.51., there exists $\mathcal{A}^{-1}$ because of

$$1 \; + \; 2a_k e_{k-1}^T A^{-1} e_k \; > \; 1 \; .$$

Thus, we find

$$\mathcal{A}^{-1} = A^{-1} - \frac{2a_k}{1 + 2a_k e_{k-1}^T A^{-1} e_k} A^{-1} e_k e_{k-1}^T A^{-1} \quad < \quad A^{-1}.$$

The inverse $\mathcal{A}^{-1}$ has the following remarkable sign pattern of
its entries

$$\mathcal{A}^{-1} \; \widehat{=} \; \begin{pmatrix} + & - & - & - & - & - & - & + \\ + & - & + & - & - & - & - & + \\ - & - & + & & & & & + \\ \vdots & \vdots & + & & & & & \vdots \\ - & - & + & - & - & - & - & + \end{pmatrix} \begin{matrix} \\ - k-1 \\ \\ \\ \end{matrix} \qquad\qquad (4.31)$$

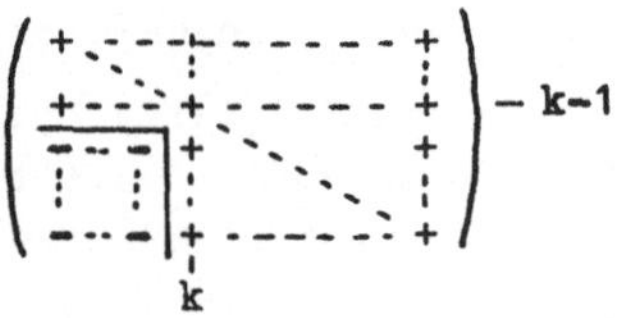

The proof is left to the reader.

If there are two positive off-diagonal entries in $\mathcal{A}$, say $a_k > 0$
and $a_j > 0$, then the inverse $\mathcal{A}^{-1}$ exists and we have the following
sign pattern

$$\mathcal{A}^{-1} \; \widehat{=} \; \begin{pmatrix} + & - & - & - & - & - & - & + \\ + & + & + & - & - & - & - & + \\ - & - & + & & & & & + \\ - & - & + & + & + & - & - & + \\ + & + & - & - & + & & & + \\ + & + & - & - & + & - & - & + \end{pmatrix} \begin{matrix} - k-1 \\ \\ - j-1 \\ \end{matrix} .$$

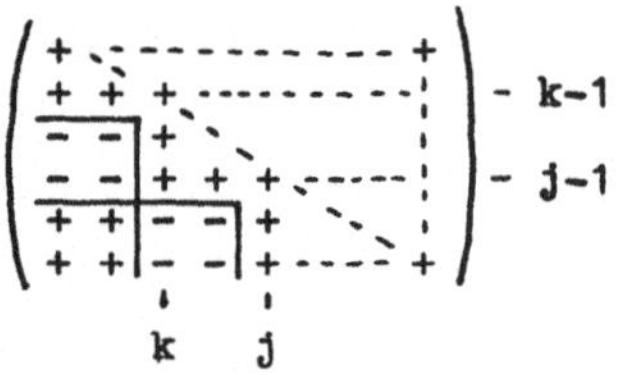

Continuing we reach another typical situation.

(b) Let the M-matrix (4.29) be changed into $\mathcal{A}$ where

$$\mathcal{A} = \begin{pmatrix} c_1 & -b_1 & & & \\ a_2 & c_2 & -b_2 & & \\ & \ddots & \ddots & \ddots & \\ & & & a_n & c_n \end{pmatrix} . \qquad\qquad (4.32)$$

That is, all the entries of the lower codiagonal of $\mathcal{A}$ are positive. Then $\mathcal{A}^{-1}$ exists which can be proved using formula (3.12)
to show that all leading principal minors $\mathcal{A}_k$ of $\mathcal{A}$ are positive.
Then the sign pattern of the entries of $\mathcal{A}^{-1}$ is the following

$$\mathcal{A}^{-1} \; \widehat{=} \; \begin{pmatrix} + & + & + & - & - & - & - & + \\ - & + & + & & & & & \vdots \\ + & - & + & & & & & \\ \vdots & & & \ddots & & & & \vdots \\ (-1)^{n+1} & - & - & - & - & + & - & + \end{pmatrix} . \qquad\qquad (4.33)$$

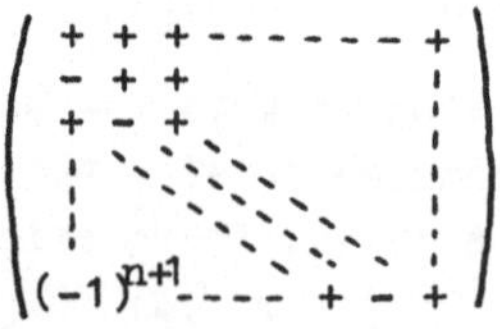

For example, let $\mathcal{A} = \mathrm{tridiag}(1,2,-1)$. Thus, we have

$n=3$ $\qquad\qquad\qquad$ $n=4$

$$\mathcal{A}^{-1} = \frac{1}{12} \begin{pmatrix} 5 & 2 & 1 \\ -2 & 4 & 2 \\ 1 & -2 & 5 \end{pmatrix}, \qquad \mathcal{A}^{-1} = \frac{1}{29} \begin{pmatrix} 12 & 5 & 2 & 1 \\ -5 & 10 & 4 & 2 \\ 2 & -4 & 10 & 5 \\ -1 & 2 & -5 & 12 \end{pmatrix},$$

$n=5$

$$\mathcal{A}^{-1} = \frac{1}{70} \begin{pmatrix} 29 & 12 & 5 & 2 & 1 \\ -12 & 24 & 10 & 4 & 2 \\ 5 & -10 & 25 & 10 & 5 \\ -2 & 4 & -10 & 24 & 12 \\ 1 & -2 & 5 & -12 & 29 \end{pmatrix},$$

and it is easily seen what the further development is.

(c) Here we change all the signs of the off-diagonal entries of A such that $\mathcal{A} \geqslant 0$. For this we introduce the diagonal matrix S given by

$$S = \operatorname{diag}(1,-1,1,-1,\ldots,(-1)^{n+1}). \qquad (4.34)$$

We see that $S = S^{-1}$. It follows that $\mathcal{A} = S^{-1}AS = SAS \geqslant 0$ and from $A^{-1} > 0$ we find

$$\mathcal{A}^{-1} = SA^{-1}S \quad \hat{=} \quad \begin{vmatrix} + & - & + & - & \cdots\cdots & (-1)^{n+1} \\ - & + & - & + & & \\ + & - & + & - & & \\ - & + & - & + & & \\ \vdots & & & & \ddots & \\ (-1)^{n+1} & \cdots\cdots\cdots & & & & + \end{vmatrix}. \qquad (4.35)$$

Thus, $\mathcal{A}^{-1}$ has a chess-board sign pattern of its entries with positive entries on the main diagonal.

Now we return to Approximation 4.1.a written as $Ay = f$ with A and f from (4.17). If there exists at least one i with $|\gamma_i| > 1$, then the M-matrix property of A is lost and it becomes clear from (a) and (b) that $A^{-1} \geqslant 0$ is impossible.

Let now $|\gamma_i| > 1$ for $i=1,\ldots,n-1$, then the striking feature of (b) shows that

$$y = A^{-1}f = u_0 A^{-1}e_1 + u_1 A^{-1}e_{n+1} = u_0 v_1 + u_1 v_{n+1},$$

is oscillating because the signs of the components of $v_1 = A^{-1}e_1$ or of $v_{n+1} = A^{-1}e_{n+1}$ are alternating, while that of v_{n+1} or v_1 are constant.

To overcome the difficulties described, we consider another FDM approximation to the Problems 4.1. and 4.1.$'$.

Approximation 4.1.b (method of upwinded differences)

Let $b(x) = b^+(x) + b^-(x)$, where $b^+ = (b + |b|)/2$ and $b^- = (b - |b|)/2$. On a uniform grid $\overline{\omega}_h$, the upwinded difference equation system is given by

$$y_0 = u_0 \, ,$$
$$- D_+D_-y_i + b_i^- D_+y_i + b_i^+ D_-y_i = 0 \, , \qquad i=1,\dots,n-1 \, ,$$
$$y_n = u_1 \, ,$$

where $b_i^\pm = b^\pm(x_i)$. The approximation is of first order accuracy.

Let $\gamma_i^+ = hb_i^+ \geqslant 0$ and $\gamma_i^- = -hb_i^- \geqslant 0$ for $i=1,..,n-1$. Thus, we have $\gamma_i^+ + \gamma_i^- = h\,|b_i|$.

Then the tridiagonal matrix A of the difference equation system $Ay = f = u_0 e_1 + u_1 e_{n+1}$ is given by

$$A = \begin{pmatrix}
1 & 0 & & & & \\
-(1+\gamma_1^+) & 2+h\,|b_1| & -(1+\gamma_1^-) & & & \\
& -(1+\gamma_2^+) & 2+h\,|b_2| & -(1+\gamma_2^-) & & \\
& & \ddots & \ddots & \ddots & \\
& & & -(1+\gamma_{n-1}^+) & 2+h\,|b_{n-1}| & -(1+\gamma_{n-1}^-) \\
& & & & 0 & 1
\end{pmatrix} . \quad (4.36)$$

We can readily see that A is an L-matrix for each $h > 0$. With the permutation matrix P defined in (4.19), we transform the matrix A into $A' = PAP^T$ of the form (3.16). Then, for any $h > 0$, the submatrix A'_{22} of A' is an irreducible nonsingular M-matrix, which implies that A' is a nonsingular M-matrix and such is also A. Further, we observe that the assumptions of Theorem 3.21. hold and thus Approximation 4.1.b satisfies the discrete maximum principle (4.22). Moreover, the solution $y = A^{-1}f$ is strongly monotone, that is, (4.23) holds if $u_0 < u_1$ and the converse if $u_0 > u_1$. This follows from (4.25), where

$$\alpha_i = \frac{1 + \gamma_i^+}{2 + \gamma_i^- + \gamma_i^+} > 0, \qquad \beta_i = \frac{1 + \gamma_i^-}{2 + \gamma_i^- + \gamma_i^+} > 0 \, , \qquad (4.37)$$

with $\alpha_i + \beta_i = 1$.

It is not difficult to see that y shows the same behaviour as indicated in Figure 4.5. For instance, let $u_0 < u_1$ and $b(x) > 0$, $x \in \Omega$. Then $\gamma_i^- = 0$ for $i=1,..,n-1$ and we find

$$y_i = \frac{1 + r_i^+}{2 + r_i^+} \, y_{i-1} + \frac{1}{2 + r_i^+} \, y_{i+1} = \frac{y_{i-1} + y_{i+1}}{2} -$$

$$- \frac{r_i^+}{2(2 + r_i^+)} (y_{i+1} - y_{i-1}) < \frac{y_{i-1} + y_{i+1}}{2} \quad , \quad i=1,\ldots,n-1.$$

Thus y is strongly convex.

Another type of FDM approximations to Problem 4.1. arises from introducing artificial diffusion or artificial viscosity, that is, the addition of an extra term of order h to the coefficient of u'', see [7,37]. This method is preferably used to approximate boundary layer problems, as will be seen later on.

<u>Approximation 4.1.c</u> (method of artificial diffusion)

Let $\varphi(x) \geq 0$, $x \in \Omega$ be a continuous function of order h. Then the artificial diffusion approximation to Problem 4.1., on a uniform grid $\overline{\omega}_h$, is defined by

$$y_o = u_o \, ,$$
$$- (1 + \varphi_i) D_+D_- y_i + b_i D_o y_i = 0 \, , \qquad i=1,\ldots,n-1,$$
$$y_n = u_1 \, ,$$

where $\varphi_i = \varphi(x_i)$.

We build the matrix form Ay = f of the system of difference equations, where A has the same tridiagonal structure as A in the Approximations 4.1.a,b and $f = u_o e_1 + u_1 e_{n+1}$. From the above approximation type we obtain the following difference equations incorporated into Ay = f.

$$-(1 + \varphi_i + \frac{hb_i}{2})y_{i-1} + 2(1 + \varphi_i)y_i - (1 + \varphi_i - \frac{hb_i}{2})y_{i+1} = 0, \quad (4.38)$$

for $i=1,\ldots,n-1$.

Now it is easily seen that $Ae \geq 0, \neq 0$ for any $h > 0$. Furthermore, A is an L-matrix if and only if

$$1 + \varphi_i \geq \frac{h|b_i|}{2} \qquad \text{for } i=1,\ldots,n-1. \quad (4.39)$$

We remark that A is a nonsingular M-matrix under conditions (4.39). To guarantee the boundary maximum principle, we require from now on the stronger conditions

$$1 + \varphi_i > \frac{h|b_i|}{2} \qquad \text{for } i=1,\ldots,n-1. \quad (4.40)$$

Then, by the same arguments used in Approximations 4.1.a,b , the method of artificial diffusion satisfies the discrete boundary maximum principle (4.22). To show the property (4.23), we consider the representation (4.25), where

$$\alpha_i = \frac{1 + \varphi_i + hb_i/2}{2(1 + \varphi_i)} > 0, \quad \beta_i = \frac{1 + \varphi_i - hb_i/2}{2(1 + \varphi_i)} > 0, \qquad (4.41)$$

and $\alpha_i + \beta_i = 1$ for $i=1,..,n-1$.

From this we have (4.27) and $y_i \neq y_{i+1}$, $i=0,..,n-1$ follows eventually from $u_o \neq u_1$.

Suppose now for definiteness $u_o < u_1$. Then y is strongly convex if $b(x) > 0$, $x \in \Omega$ and strongly concave if $b(x) < 0$, $x \in \Omega$, see also Figure 4.5. These conclusions follow from (4.25), where (4.41) is used. For example

$$y_i = \frac{y_{i-1} + y_{i-1}}{2} - \frac{hb_i}{4(1 + \varphi_i)} (y_{i+1} - y_{i-1}) < \frac{y_{i-1} + y_{i+1}}{2} ,$$

for $i=1,..,n-1$ if $b(x) > 0$, $x \in \Omega$. Thus, y is strongly convex.

Many specifications of $\varphi(x)$ can be found in the literatur, see [7,49].

It is interesting to note that the choice $\varphi(x) = h|b(x)|/2$, which satisfies (4.40), reproduces the method of upwinded differences discussed in Approximation 4.1.b. Evidently, we have

$$2(1 + \varphi_i) = 2 + h|b_i| = 2 - hb_i^- + hb_i^+ = 2 + \gamma_i^- + \gamma_i^+ ,$$
$$1 + \varphi_i + hb_i/2 = 1 + hb_i^+ = 1 + \gamma_i^+ ,$$
$$1 + \varphi_i - hb_i/2 = 1 - hb_i^- = 1 + \gamma_i^- .$$

It is not our goal to review all of the possible choices of $\varphi(x)$ known from the literature.

Here, we again turn to simple singularly perturbed problems. We shall vary Problem 4.1.' and consider the following model problem.

Problem 4.1.''

$$Lu = -\varepsilon u'' + b(x) u' = 0 , \quad x \in \Omega ,$$
$$u(0) = u_o , \quad u(1) = u_1 .$$

Suppose that $\varepsilon > 0$ is a small parameter.

We see that Problem 4.1.'' has the same properties as Problem 4.1. In particular, the boundary maximum principle (4.14) holds. But in comparison with Problem 4.1.', inner layers may occur.

Let us do such an example as given in [7].

Example 4.1.1.

$$- \varepsilon u'' - 2x u' = 0 , \quad x \in (-1,1),$$
$$u(-1) = -1 , \quad u(1) = 2 .$$

The solution

$$u(x) = 3 \frac{\int_{-1}^{x} \exp(- \frac{t^2}{\varepsilon}) dt}{\int_{-1}^{+1} \exp(- \frac{t^2}{\varepsilon}) dt} - 1 ,$$

which exhibits an inner layer near x=0, is depicted in Figure 4.6.

Fig.4.6.

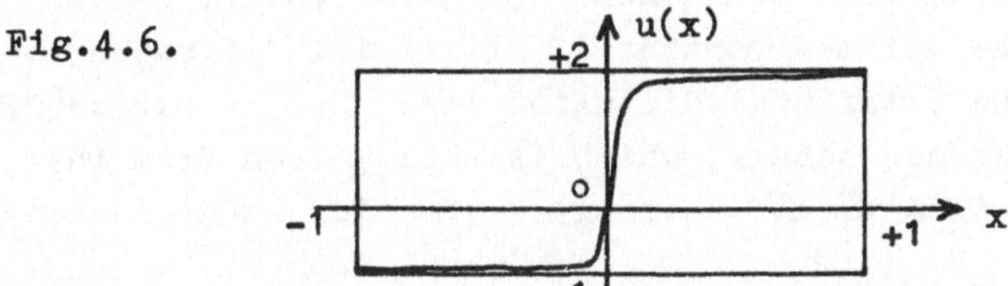

According to Problem 4.1.'' the stability condition (4.40) now takes
the form

$$\varepsilon + \varsigma_i > \frac{h|b_i|}{2} , \quad i=1,...,n-1 . \qquad (4.40')$$

The introduction of such a portion of artificial diffusion $\varsigma(x)$,
which is not adequate to the problem under consideration leads to
diffusion of the layer. For instance, this is the case employing an
upwinded scheme by $\varsigma(x) = h|b(x)|/2$. This scheme is stable for any
h > 0 but models the boundary layer badly, see [7]. In fact, upwinded
schemes are good for approximating the behaviour of the solution
outside the boundary layer. But, they exhibit the undesirable property that the error may increase significantly in the boundary
layer region as h decreases and grid points are placed in the boundary layer. When the step size h decreases sufficiently, the error
eventually begins to decline again. To avoid such small h, a stronger convergence criterion known as uniform convergence has been introduced, see [7,33].
That is

$$|y_i - u(x_i)| \leq C h^p, \qquad (4.42)$$

for i=1,..,n-1, where p > 0 and C = const is independent of both
h and ε .

Uniform convergence is sufficient to guarantee that the problem can
be solved accurately on a coarse grid and that the boundary layer
will be resolved in the right behaviour.

It is well known that classical methods do not satisfy the criterion
of uniform convergence. The criterion of uniform convergence was
first introduced by Il'in, see [38], who proved that his FDM approach
to Problem 4.1.'' was uniformly convergent of order p=1.
For generalizations and further developments we refer to [7,33,37].

Here, we briefly consider the famous Il'in scheme from our point of
view, which is defined by the choice

$$\varsigma(x) = \frac{hb(x)}{2}\left(\coth\frac{hb(x)}{2\varepsilon} - \frac{2\varepsilon}{hb(x)}\right), \qquad (4.43)$$

for the boundary layer Problem 4.1.'', assuming $|b(x)| \geqslant b_o = \text{const} > 0$,
$x \in \Omega$. It is easily seen that this choice of $\varsigma(x)$ satisfies the
condition (4.40'). The defined portion of artificial diffusion leads
to a refinement of the artificial diffusion term $\varsigma(x) = h|b(x)|/2$
of the upwinded difference scheme, which is easily seen from the
behaviour of the function $\Phi(x) = \coth(x) - 1/x$ for $x \in R^1$ depicted
in Figure 4.7.

Fig.4.7.

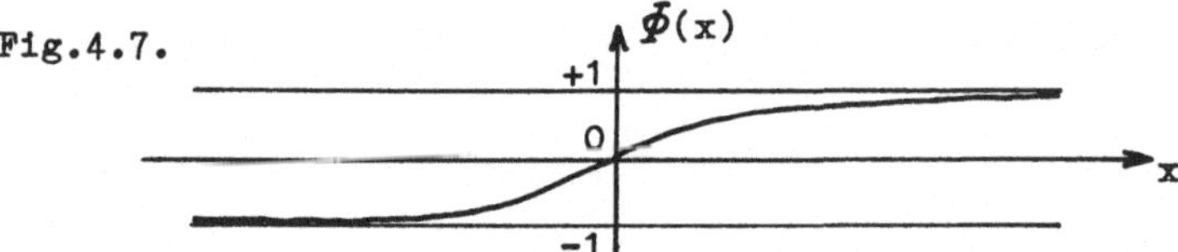

Using $\varsigma(x)$ from (4.43), the Approximation 4.1.c applied to Problem
4.1.'' now takes the form

$$y_o = u_o ,$$

$$-\varepsilon\,\varkappa\left(\frac{hb_i}{2\varepsilon}\right)D_+D_-y_i + b_iD_oy_i = 0 , \qquad i=1,..,n-1 , \quad (4.44)$$

$$y_n = u_1 ,$$

where $\varkappa(z) = z\coth(z)$.

Thus we have the following difference equations for i=1,..,n-1

$$-\left(\varkappa\left(\frac{hb_i}{2\varepsilon}\right) + \frac{hb_i}{2\varepsilon}\right)y_{i-1} + 2\varkappa\left(\frac{hb_i}{2\varepsilon}\right)y_i - \left(\varkappa\left(\frac{hb_i}{2\varepsilon}\right) - \frac{hb_i}{2\varepsilon}\right)y_{i+1} = 0, (4.45)$$

which build, with the boundary conditions of (4.44), the matrix
form Ay = f of the system of difference equations. The resulting
tridiagonal matrix A has the same structure as the matrices defined
by (4.17) and (4.36).
It is obvious that $Ae \geqslant 0$, $\neq 0$ for any h > 0.
Further, A is an L-matrix because for $|z| \geqslant hb_o/(2\varepsilon)$ the following
inequalities

$$\mathscr{x}(z) = z\coth(z) > 0,$$
$$\mathscr{x}(z) + z = z(\coth(z) + 1) > 0, \qquad (4.46)$$
$$\mathscr{x}(z) - z = z(\coth(z) - 1) > 0,$$

hold.

By the same arguments used in Approximations 4.1.a-c, we find that A
is a nonsingular M-matrix and $Ay = f$ satisfies the discrete boundary
maximum principle (4.22), which further implies (4.23).
In the representation (4.25) we have with $z_i = hb_i/(2\varepsilon)$ from (4.45)

$$\alpha_i = \frac{\mathscr{x}(z_i) + z_i}{2\,\mathscr{x}(z_i)} > 0, \qquad \beta_i = \frac{\mathscr{x}(z_i) - z_i}{2\,\mathscr{x}(z_i)} > 0, \qquad (4.47)$$

with $\alpha_i + \beta_i = 1$. Hence, (4.27) holds.
If $u_0 < u_1$ and $b(x) > 0$ for $x \in \Omega$, then y is strongly convex. This is
seen from

$$y_i = \frac{y_{i-1} + y_{i+1}}{2} - \frac{1}{2\coth(z_i)}\,(y_{i+1} - y_{i-1}) < \frac{y_{i-1} + y_{i+1}}{2},$$

for $i=1,..,n-1$.
The Il'in scheme is exact for Problem 4.1.$''$, putting $b(x) = 1$.

Now we turn to another problem.

Problem 4.2.

$$\tilde{L}u = Lu + q(x)u = \tilde{f}(x), \quad x \in \Omega,$$
$$u(0) = u_0, \quad u(1) = u_1.$$

Suppose that L is given by one of the Problems 4.1. or 4.1.$''$.
Further, let $q(x)$ and $\tilde{f}(x)$ be sufficiently smooth where we impose
additionally

$$q(x) \geqslant 0 \quad \text{for } x \in \Omega. \qquad (4.48)$$

Assuming these conditions, Problem 4.2. satisfies the following
maximum principle.
Let $\tilde{f}(x) \leqslant 0$, $x \in \Omega$, then a nonnegative maximum (a nonpositive
minimum if $\tilde{f}(x) \geqslant 0$, $x \in \Omega$) of a solution $u(x)$, if it exists, is
taken at $\partial\Omega = \{0,1\}$. The reader wishing to acquire a deeper insight
into maximum principles according to Problem 4.2. is referred to [22].
In addition, we only pick out a special maximum principle assuming
$Lu = -\varepsilon u''$, $\varepsilon = \text{const} > 0$ and $q(x) \geqslant q_0 = \text{const} > 0$, $x \in \Omega$.
Then under the assumptions $u_0 \geqslant 0$, $u_1 \geqslant 0$ and $\tilde{f}(x) \geqslant 0$, $x \in \Omega$ we have
$u(x) \geqslant 0$ for $x \in \overline{\Omega}$, see [7].
For example consider the model problem

$$Lu = -\varepsilon u'' + u = 0, \quad x \in \Omega, \qquad (4.49)$$
$$u(0) = u(1) = 1 + \exp(-1/\sqrt{\varepsilon}),$$

where $\varepsilon > 0$ is a small parameter. Problem (4.49) has the solution

$$0 \leq u(x) = \exp\left(-\frac{x}{\sqrt{\varepsilon}}\right) + \exp\left(-\frac{1-x}{\sqrt{\varepsilon}}\right) \leq 1 + \exp\left(-\frac{1}{\sqrt{\varepsilon}}\right) ,$$

for $x \in \bar{\Omega}$, which exhibits boundary layers near x=0 and x=1, see [7].
The behaviour of such a solution u(x) is depicted in Figure 4.8.

Fig.4.8.

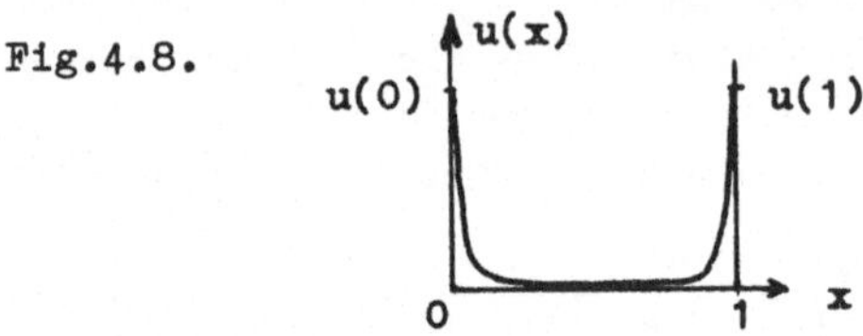

There is proof in [7,33] that uniform convergence can be obtained
only if the difference scheme is exponentially fitted, i.e.,if the
coefficients in the difference equations contain exponential functions. However, fundamental difficulties arise when the method is
to be generalized to more dimensions and little progress has been
reported so far.

The FDM approximation to Problem 4.2., on a uniform grid $\bar{\omega}_h$, is
defined as follows.

Approximation 4.2.

$$\tilde{L}_h y_i = L_h y_i + q_i y_i = \tilde{f}_i , \quad i=1,..,n-1,$$

$$y_0 = u_0 , \quad y_n = u_1 ,$$

$q_i = q(x_i) \geq 0$, $\tilde{f}_i = \tilde{f}(x_i)$ and L_h is an FDM approximation to L,
for instance, one of the Approximations 4.1.a-c.
Let

$$A = \begin{pmatrix} 1 & 0 & & & \\ -a_1 & c_1 & -b_1 & & \\ & \ddots & \ddots & \ddots & \\ & & -a_{n-1} & c_{n-1} & -b_{n-1} \\ & & & 0 & 1 \end{pmatrix} \tag{4.50}$$

be a nonsingular tridiagonal M-matrix which corresponds to the
partial problem

$$L_h y_i = \tilde{f}_i , \quad i=1,..,n-1,$$
$$y_0 = u_0 , \quad y_n = u_1, \tag{4.51}$$

and satisfies $Ae = (1,0,..,0,1)^T \geq 0$.
Then the matrix form of Approximation 4.2. is given by

$$\tilde{A} y = f , \tag{4.52}$$

where $\tilde{A} = A + Q$, $Q = \text{diag}(0,q_1,...,q_{n-1},0) \geq 0$ and the right-
hand side vector $f=(u_0,\tilde{f}_1,...,\tilde{f}_{n-1},u_1)^T$.

By virtue of Theorem 3.12., $\tilde{A}$ is a nonsingular M-matrix and

$$0 \leq \tilde{A}^{-1} \leq A^{-1}.$$

Hence, $f \geq 0$ in (4.52) immediately implies

$$0 \leq y = \tilde{A}^{-1}f \leq A^{-1}f,$$

where $A^{-1}f$ is the solution of problem (4.51).

Let us look more closely at the difference equation system (4.52), applying the region maximum principle, see Section 3.4.2. First, let $q(x) > 0$ for $x \in \Omega$. Then A is a strongly row diagonally dominant M-matrix, that is

$$\tilde{A}e = (1, q_1, \ldots, q_{n-1}, 1)^T > 0.$$

Now we define $B = C = \mathrm{diag}(1, q_1, \ldots, q_{n-1}, 1)$, $v = B^{-1}f$, where

$$v = (u_0, \tilde{f}_1/q_1, \ldots, \tilde{f}_{n-1}/q_{n-1}, u_1)^T,$$

and rewrite the system (4.52) as

$$\tilde{A}y = Bv, \tag{4.53}$$

which has the form of the equation system (3.28), putting $w = 0$ and $\tau > 0$. It has been found that (4.53) satisfies the assumptions of Theorem 3.24. Thus, the components of the solution $y = (y_0, \ldots, y_n)^T$ of (4.52) are bounded as follows

$$\min \left\{ u_0, \frac{\tilde{f}_1}{q_1}, \ldots, \frac{\tilde{f}_{n-1}}{q_{n-1}}, u_1 \right\} \leq y_i \leq \max \left\{ u_0, \frac{\tilde{f}_1}{q_1}, \ldots, \frac{\tilde{f}_{n-1}}{q_{n-1}}, u_1 \right\}, \tag{4.54}$$

for $i = 0, \ldots, n$.

The inequalities (4.54) also reflect the maximum principle of Problem 4.2. For instance, let $\tilde{f}(x) \leq 0$, $x \in \Omega$ and $\max\{u_0, u_1\} \geq 0$. Then our conclusion from (4.54) is

$$y_i \leq \max \{u_0, u_1\}, \quad i = 0, \ldots, n,$$

and conversely, $y_i \geq \min \{u_0, u_1\}$ if $\tilde{f}(x) \geq 0$, $x \in \Omega$ and $\min \{u_0, u_1\} \leq 0$.

Second, let $q(x) \geq 0$, $x \in \Omega$, for instance $q(x) \equiv 0$. Then $\tilde{A}$ is not strongly row diagonally dominant. To show that in this case the maximum principle also holds, we consider the following perturbed system (4.52)

$$(\tilde{A} + \delta I)y^\delta = f, \tag{4.55}$$

where $\delta > 0$ is a constant. By Theorem 3.12., the matrix $\tilde{A} + \delta I$ is a nonsingular M-matrix for any $\delta > 0$ and it is strongly row diagonally dominant, i.e. $(\tilde{A} + \delta I)e > 0$ for $\delta > 0$.

Let
$$(\tilde{A} + \delta I) e = (1+\delta, r_1, \ldots, r_{n-1}, 1+\delta)^T > 0,$$

we define $B = C = \mathrm{diag}(1+\delta, r_1, \ldots, r_{n-1}, 1+\delta) \geqslant 0$ and rewrite the system (4.55) as follows
$$(\tilde{A} + \delta I) y^\delta = B v , \qquad (4.56)$$

where
$$v = B^{-1} f = \left(\frac{u_0}{1+\delta}, \frac{\tilde{f}_1}{r_1}, \cdots, \frac{\tilde{f}_{n-1}}{r_{n-1}}, \frac{u_1}{1+\delta} \right)^T .$$

Putting $w=0$, $\tau > 0$, the system (4.56) is of the form (3.28). It is now easily seen that the conditions of Theorem 3.24. hold. Thus, the solution $y^\delta = (y_0^\delta, y_1^\delta, \ldots, y_n^\delta)^T$ of (4.56) satisfies
$$\min \left\{ \frac{u_0}{1+\delta}, \frac{\tilde{f}_1}{r_1}, \ldots, \frac{\tilde{f}_{n-1}}{r_{n-1}}, \frac{u_1}{1+\delta} \right\} \leqslant y_i^\delta \leqslant \max \left\{ \frac{u_0}{1+\delta}, \frac{\tilde{f}_1}{r_1}, \ldots, \frac{\tilde{f}_{n-1}}{r_{n-1}}, \frac{u_1}{1+\delta} \right\}, \quad (4.57)$$

for $i=0,\ldots,n$.

Now we deduce from (4.57) the discrete analogue to the maximum principle of Problem 4.2. For this let $\tilde{f}(x) \leqslant 0$, $x \in \Omega$ and $\max\{u_0, u_1\} \geqslant 0$. Then we have
$$y_i^\delta \leqslant \max \left\{ \frac{u_0}{1+\delta}, \frac{u_1}{1+\delta} \right\} \qquad \text{for } i=0,1,\ldots,n.$$

Let $\delta \longrightarrow +0$, then, by continuity arguments, it follows that $y^\delta \longrightarrow y$, where y is the solution of problem (4.52). Hence
$$y_i \leqslant \max \{u_0, u_1\} \qquad \text{for } i=0,1,\ldots,n.$$

Conversely, let $\tilde{f}(x) \geqslant 0$, $x \in \Omega$ and $\min\{u_0, u_1\} \leqslant 0$, then we get
$$\min \{u_0, u_1\} \leqslant y_i \qquad \text{for } i=0,1,\ldots,n.$$

We proceed to briefly consider the specified Problem 4.2., assuming $Lu = -\varepsilon u''$ with $\varepsilon > 0$ and $q(x) \geqslant q_0 > 0$ for $x \in \Omega$. The FDM approximation of this partial problem is defined as follows
$$y_0 = u_0 ,$$
$$- \varepsilon D_+ D_- y_i + q_i y_i = \tilde{f}_i , \quad i=1,\ldots,n-1, \qquad (4.58)$$
$$y_n = u_1 .$$

Let the matrix form of the difference equation system (4.58) be $\tilde{A} y = f$. Then it is easily seen that A is a strongly row diagonally dominant M-matrix, that is
$$\tilde{A} e = (1, q_1, \ldots, q_{n-1}, 1)^T > 0,$$

and the right-hand side vector f is defined as in (4.52). Let
$B = C = \text{diag}(1, q_1, \ldots, q_{n-1}, 1) \geqslant 0$. Then we rewrite $\tilde{A}y = f$ as $\tilde{A}y = Bv$,
where $v = B^{-1}f$. Applying Theorem 3.24., we immediately deduce the
following solution bounds for the solution y of (4.58)

$$\min\left\{u_o, \frac{\tilde{f}_1}{q_1}, \ldots, \frac{\tilde{f}_{n-1}}{q_{n-1}}, u_1\right\} \leqslant y_i \leqslant \max\left\{u_o, \frac{\tilde{f}_1}{q_1}, \ldots, \frac{\tilde{f}_{n-1}}{q_{n-1}}, u_1\right\}, \qquad (4.59)$$

for $i = 0, \ldots, n$.

Thus, assuming $\tilde{f}(x) \geqslant 0$, $x \in \Omega$ and $\min\{u_o, u_1\} \geqslant 0$, we conclude from
(4.59) that $y \geqslant 0$ which reflects the above mentioned maximum prin-
ciple.

Moreover, let $\tilde{f}(x) = 0$, $x \in \Omega$, then we have from (4.59) the enclosure of the solution y of (4.58)

$$\min\left\{u_o, 0, u_1\right\} \leqslant y_i \leqslant \max\left\{u_o, 0, u_1\right\} \qquad \text{for } i = 0, \ldots, n.$$

Let us note that the above used methods of investigating qualitative
properties of FDM approximations to Problems 4.1. and 4.2. are also
applicable to difference schemes for problems stated in divergence
form like the following

$$\tilde{L}u = -(p(x)u' + b(x)u)' + q(x)u = \tilde{f}(x), \qquad x \in \Omega,$$

$$u(0) = u_o, \quad u(1) = u_1.$$

The reason is that the matrix form of the difference equation systems
involves nonsingular M-matrices.

It should be pointed out that much more sophisticated FDM approximations are of need if the problems under consideration involve boundary or interior layers and the schemes are to be uniformly convergent. A first impression gives the famous Il'in scheme for the Problem 4.1.''. Classes of uniform convergent difference schemes for
one-dimensional singularly perturbed problems of the type

$$-\varepsilon u'' + b(x)u' + q(x)u = \tilde{f}(x), \qquad x \in \Omega,$$

$$u(0) = u_o, \quad u(1) = u_1,$$

under the restriction $|b(x)| \geqslant b_o > 0$, $x \in \Omega$, have been described
in [7].

For uniform convergent difference schemes for singularly perturbed
problems which involve turning points, that is, the coefficient
$b(x)$ has zeros in Ω, see also Example 4.1.1., we refer to [33]
and elsewhere.

4.3.2. Difference approximations to two-dimensional elliptic boundary value problems

Our object is now to consider FDM approximations to model problems of convection-diffusion type over bounded two-dimensional regions Ω . The first model problem under consideration is the following.

Problem 4.3.

$$Lu \quad = \quad -\,\Delta u \quad + \quad (b(x))^T \nabla u \quad = \quad 0, \quad x \in \Omega ,$$

$$u \quad = \quad g(x), \quad x \in \partial\Omega .$$

Suppose that the boundary $\partial\Omega$ of $\Omega \subset R^2$ and the functions $b(x) = (b_1(x),b_2(x))^T$, $g(x)$ are sufficiently smooth such that Problem 4.3. has a unique solution $u(x) \in C^2(\Omega) \cap C(\bar{\Omega})$. Then, by the boundary maximum principle, both $\max_{z \in \bar{\Omega}} u(z)$ and $\min_{z \in \bar{\Omega}} u(z)$ for $u \neq$ const are taken on at the boundary $\partial\Omega$, see [22].
Hence

$$\min_{z \in \partial\Omega} u(z) \quad \leqslant \quad u(x) \quad \leqslant \quad \max_{z \in \partial\Omega} u(z), \quad x \in \bar{\Omega} . \qquad (4.60)$$

Letting $b(x) \equiv 0$, $x \in \Omega$, we get the simplest example of an elliptic equation, Laplace's equation in R^2, i.e. $-\Delta u = 0$. It is known from elementary function theory that all continuous solutions of $-\Delta u = 0$, the harmonic functions, are smooth and in fact are analytic. It is common knowledge that harmonic functions obey the maximum principle (4.60) and moreover satisfy the mean value property

$$u(x) \quad = \quad \frac{1}{2\pi r} \int_{\partial \mathcal{K}_r(x)} u(s)\,ds \qquad (4.61)$$

for each disc $\mathcal{K}_r(x) = \{y: \|y - x\|_2 < r\} \subset \Omega$, see [11].

For $\|b(x)\| \gg 1$, Problem 4.3. may involve boundary layers which are usually of several types on certain parts of $\partial\Omega$.
Let, for simplicity, $b(x) = b = (b_1,b_2)^T/\varepsilon$, where b_1,b_2 are constants and $\varepsilon > 0$ is a small parameter. Then we have the following simple singularly perturbed problem.

Problem 4.3.$'$

$$Lu \quad = \quad -\varepsilon\Delta u \quad + \quad b_1\frac{\partial u}{\partial x_1} \quad + \quad b_2\frac{\partial u}{\partial x_2} \quad = \quad 0, \quad x \in \Omega ,$$

$$u \quad = \quad g(x), \quad x \in \partial\Omega .$$

It is thus clear that the boundary maximum principle (4.60) carries over to Problem 4.3.$'$.

Let us now begin by considering FDM approximations to Problem 4.3.
on rectangular grids $\overline{\omega}_h$. For brevity, we assume Ω of rectangular
shape with parallel sides to the coordinate axis. We remark that
FDM approximations on rectangular grids over arbitrary regions $\Omega \subset R^2$
are discussed in [11,23] among others.

Suppose now for definiteness $\Omega = (0,1)^2$. Further, let $\overline{\omega}_h$, $h = (h_1, h_2)$
with $h_k = 1/n_k$, be a uniform rectangular grid defined by

$$\overline{\omega}_h = \left\{ x_{ij} = (ih_1, jh_2), \ i=0,..,n_1, \ j=0,..,n_2 \right\} \setminus \left\{ (0,0),(0,1),(1,0),(1,1) \right\}$$

$$\overline{\omega}_h = \omega_h + \gamma_h , \tag{4.62}$$

where $\omega_h = \left\{ x_{ij} \in \Omega \right\}$, $\gamma_h = \left\{ x_{ij} \in \partial\Omega \right\}$.

Denote by y_{ij} the approximate values of the solution of Problem 4.3.
at the grid points x_{ij}, i.e. $y_{ij} \approx u(x_{ij})$ for all $x_{ij} \in \overline{\omega}_h$.

Using the five-point difference star $\mathcal{S}(i,j) = \left\{ x_{ij}, x_{i\pm1j}, x_{ij\pm1} \right\}$,
we consider the following FDM approximation to Problem 4.3.

Approximation 4.3. (method of artificial diffusion)

Let $\mathcal{S}(x) = (\mathcal{S}_1(x), \mathcal{S}_2(x))^T \geq 0$, $x \in \Omega$ be continuous and of order
h, i.e. $\mathcal{S}_k(x) = O(h_k)$, $k=1,2$.

The artificial diffusion scheme, on the uniform grid (4.62), is then
defined as

$$\sum_{k=1}^{2} \left(-(1+\mathcal{S}_{ij}^k)D_+^k D_-^k + b_{ij}^k D_o^k \right) y_{ij} = 0, \qquad x_{ij} \in \omega_h,$$

$$y_{ij} = g_{ij}, \quad x_{ij} \in \gamma_h,$$

where D_+^k, D_-^k, D_o^k are the difference operators with respect to the
x_k-axis directions, $\mathcal{S}_{ij}^k = \mathcal{S}_k(x_{ij})$, $b_{ij}^k = b_k(x_{ij})$ and $g_{ij} = g(x_{ij})$.
The scheme is at least of first order accuracy.

Denote by y the vector with components y_{ij}, $0 \leq i \leq n_1$, $0 \leq j \leq n_2$,
listed row by row. Then the system of difference equations, the
boundary values $y_{ij} = g_{ij}$ as single equations incorporated, can be
written in matrix form

$$A y = f ,$$

where the right-hand side f is made up of components 0 and g_{ij}.

To analyse the main properties of Approximation 4.3., we rewrite
the five-point difference equations as follows

$$-\frac{1}{h_1^2}\left(1 + \varphi_{ij}^1 + h_1 b_{ij}^1/2\right) y_{i-1j} \quad + \quad 2\left(\frac{1+\varphi_{ij}^1}{h_1^2} + \frac{1+\varphi_{ij}^2}{h_2^2}\right) y_{ij} -$$

$$-\frac{1}{h_1^2}\left(1 + \varphi_{ij}^1 - h_1 b_{ij}^1/2\right) y_{i+1j} \quad - \quad \frac{1}{h_2^2}\left(1 + \varphi_{ij}^2 + h_2 b_{ij}^2/2\right) y_{ij-1}$$

$$-\frac{1}{h_2^2}\left(1 + \varphi_{ij}^2 - h_2 b_{ij}^2/2\right) y_{ij+1} \quad = \quad 0 , \qquad\qquad (4.63)$$

which hold for each $x_{ij} \in \omega_h$.

It is now readily verified that A satisfies

$$A e \geqslant 0, \neq 0 .$$

From (4.63) it is seen that the diagonal entries of A are for any choice $\varphi(x) \geqslant 0$ positive and the off-diagonal entries of A are nonpositive if and only if

$$1 + \varphi_k(x) \geqslant h_k |b_k(x)| /2 , \qquad x \in \Omega , \quad k=1,2$$

holds.
This condition needs sharpening

$$1 + \varphi_k(x) > h_k |b_k(x)| /2 , \qquad x \in \Omega , \quad k=1,2, \quad (4.64)$$

as will be seen later on.

With a well defined permutation matrix P we now transform the system $Ay = f$ into the system $A'y' = f'$, which is exactly of the form (3.18), putting

$$A' = PAP^T = \begin{pmatrix} I & 0 \\ A'_{21} & A'_{22} \end{pmatrix}, \quad y' = Py , \quad f' = Pf .$$

The first $2(n_1-1)+2(n_2-1)$ components of f' are the boundary values $g_{ij} = g(x_{ij})$, $x_{ij} \in \gamma_h$.
We have

$$(A'_{21} \; A'_{22}) e = 0 , \qquad A'_{21} \leqslant 0 .$$

Assuming now (4.64), we get $(A'_{21})^T e'' < 0'$ where e'' and $0'$ are vectors all of ones and zeros, respectively. Moreover, the assumption implies that A'_{22} is an irreducible weakly row diagonally dominant M-matrix. Thus, A' is a nonsingular M-matrix and so is A.

Summing up, it has been found that the conditions of Theorem 3.21.
hold under the assumption (4.64), and thus Approximation 4.3. satisfies the discrete boundary maximum principle defined in Section
3.4.1. in analogy to the continuous maximum principle (4.60).
That is,

$$\min_{x_{kl} \in \gamma_h} g_{kl} \;\leq\; y_{ij} \;\leq\; \max_{x_{kl} \in \gamma_h} g_{kl} \qquad \text{for each } x_{ij} \in \overline{\omega}_h. \quad (4.65)$$

The solution y of Approximation 4.3. avoids local maxima and minima
over ω_h. In order to make this property more transparent, we rewrite the difference equations (4.63) as

$$y_{ij} = \alpha_{ij} y_{i-1j} + \beta_{ij} y_{i+1j} + \gamma_{ij} y_{ij-1} + \delta_{ij} y_{ij+1}, \quad (4.66)$$

for each $x_{ij} \in \omega_h$, where $\alpha_{ij} > 0$, $\beta_{ij} > 0$, $\gamma_{ij} > 0$, $\delta_{ij} > 0$.
The formulas for this coefficients are easily available from (4.63)
which then satisfy

$$\alpha_{ij} + \beta_{ij} + \gamma_{ij} + \delta_{ij} = 1 .$$

Thus, any solution component y_{ij} for $x_{ij} \in \omega_h$ is a convex linear
combination of its neighbouring solution components $y_{i\pm1j}$, $y_{ij\pm1}$.
Hence, we have

$$\min \left\{ y_{i\pm1j}, \; y_{ij\pm1} \right\} \;\leq\; y_{ij} \;\leq\; \max \left\{ y_{i\pm1j}, \; y_{ij\pm1} \right\}, \qquad (4.67)$$

for each $x_{ij} \in \omega_h$.

It is now possible to investigate, in analogy to the one-dimensional
problems, which further conditions on b(x) and $\varrho(x)$ lead to locally
convex or concave behaviour of the approximate solution y, that is,
in which cases we have, over the difference star $\mathscr{S}(i,j)$,

$$y_{ij} \leq \overline{y}_{ij} \qquad \text{or} \qquad y_{ij} \geq \overline{y}_{ij} ,$$

where $\overline{y}_{ij} = (y_{i-1j} + y_{i+1j} + y_{ij-1} + y_{ij+1})/4$.

Let $b(x) \equiv 0$ and choose $\varrho(x) \equiv 0$ for $x \in \Omega$, then y is an approximation to a harmonic function. If additionally $h_1 = h_2$, this means
$\overline{\omega}_h$ is a square grid, we get

$$y_{ij} = \overline{y}_{ij} \qquad \text{for each } x_{ij} \in \omega_h, \qquad (4.68)$$

which is a discrete analogue to the mean value property (4.61) of
harmonic functions, see [11].

Next, let us briefly discuss the two-dimensional upwinded difference
scheme. It is generated by the choice

$$\varphi(x) \;=\; (\; h_1|b_1(x)|/2 \;,\; h_2|b_2(x)|/2 \;) \;,\qquad x \in \Omega \;,$$

which satisfies condition (4.64) and leads to absolute **stable** FDM
approximations to Problem 4.3. satisfying the discrete boundary
maximum principle.

According to Problem 4.3.´ the condition (4.64) changes as follows
into condition

$$\varepsilon \;+\; \varphi_k(x) \;>\; h_k|b_k(x)|/2 \;,\qquad x \in \Omega \;,\qquad k=1,2 \;. \qquad (4.69)$$

We note that it is much more difficult to construct good approximations to two-dimensional singularly perturbed problems rather than
to one-dimensional ones. This is borne out by the fact that as yet
no two-dimensional analogue to the Il´in scheme is known.

Next we shall study an FDM approximation to an elliptic problem
stated in divergence form over triangulized regions.
For this, let $\Omega \subset R^2$ be a bounded, convex and polygonal region, i.e.
$\partial\Omega$ is Lipschitzian. The elliptic boundary value problem under consideration is the following.

<u>Problem 4.4.</u>

$$Lu \;=\; -\nabla(k(x)\nabla u \;+\; b(x)\,u) \;=\; 0 \;,\qquad x \in \Omega \;,$$

$$u \;=\; g(x) \;,\qquad x \in \partial\Omega \;.$$

Suppose that $k(x)$, $b(x) = (b_1(x),b_2(x))^T$ are sufficiently smooth
with $k(x) \geq k_0 = \text{const} > 0$, $x \in \Omega$ such that Problem 4.4. has a
classical solution $u(x) \in C^2(\Omega) \cap C(\bar{\Omega})$.

To construct an FDM approximation to Problem 4.4., we first describe
the triangulation of $\bar{\Omega}$. We naturally assume that adjacent triangles share a common side and, furthermore, that the triangulation of
$\bar{\Omega}$ is of weakly acute type. The latter means that there is no obtuse-
angled triangle in the subdivision of $\bar{\Omega}$. With a given triangulation
we associate the FDM grid $\bar{\omega}_h$. Let $\bar{\omega}_h = \{ x_i, \; i=1,\ldots,n \}$. Then $\bar{\omega}_h$
is defined by the set of all vertices x_i of the triangles. Further,
let

$$\bar{\omega}_h \;=\; \omega_h \;+\; \gamma_h \;,$$

where

$$\gamma_h \;=\; \{ x_i \in \partial\Omega \;,\quad i=1,\ldots,n_1 \}$$
$$\omega_h \;=\; \{ x_i \in \Omega \;,\quad i=n_1+1,\ldots,n \} \;.$$

To derive difference equations for each $x_i \in \omega_h$, we next define
circumcentric elementary regions $\mathcal{H}_i$ with $x_i \in \overline{\mathcal{H}}_i$ as shown in
Figure 4.9.

Fig.4.9.

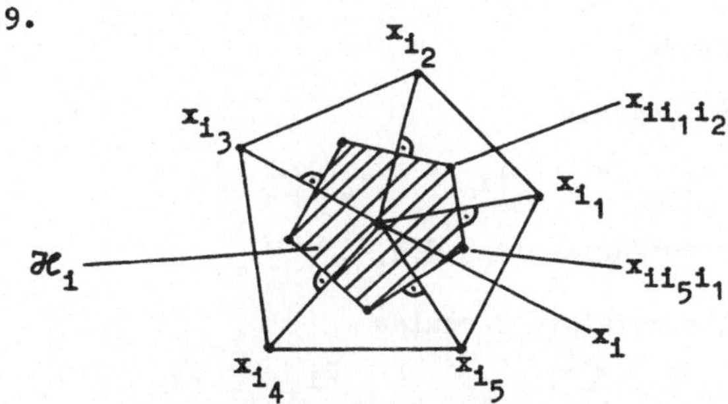

Denote by x_{ijs} the midpoint of the circumference of the triangle
with vertices x_i, x_j, x_s. Then x_{ijs} is not placed outside of this
triangle by our assumptions.
Let

$$h_{ij} = \| x_i - x_j \|_2 > 0 \quad \text{for each triangle side } \overline{x_i x_j},$$

$$l_{ij} = \| x_{ijs} - x_{jim} \|_2 \geq 0 \quad \text{for each pair of adjacent triangles.}$$

Denote by ν_i the unit outward normal of the elementary region $\mathcal{H}_i$
for each $x_i \in \omega_h$. Integrating the differential equation of Problem
4.4. over $\mathcal{H}_i$ and applying the partial integration formula (4.1)
for $v \equiv 1$, we have

$$- \int_{\partial \mathcal{H}_i} (k(x) \nabla u + b(x) u) \, ds = 0 \quad \text{for each } x_i \in \omega_h, \quad (4.70)$$

where $ds = (\cos(\nu_i, x_1\text{-axis}), \cos(\nu_i, x_2\text{-axis})) \, ds'$, see [12,13].
For sake of simplicity, we now impose a potential $\varphi(x)$, $x \in \Omega$ such
that

$$\nabla \varphi(x) = b(x) = (b_1(x), b_2(x))^T. \quad (4.71)$$

Therefore, the curve integral (4.70) takes the form

$$- \int_{\partial \mathcal{H}_i} (k(x) \frac{\partial u}{\partial \nu_i} + u \frac{\partial \varphi}{\partial \nu_i}) \, ds' = 0. \quad (4.72)$$

Denote by y_i the approximate values of the solution $u(x)$ of Problem
4.4. at grid points $x_i \in \overline{\omega}_h$, where

$$y_i = g(x_i) = g_i \quad \text{for } x_i \in \gamma_h.$$

Further, for ease of notation, let $\mathscr{S}(i)$ be the difference star defined by

$$\mathscr{S}(i) = \left\{ x_{i_m} \in \overline{\omega}_h \text{ where } x_{i_m} \neq x_i \text{ are vertices of triangles with one vertex } x_i \right\},$$

see also Figure 4.9. and let

$$x_{ii_m} = (x_i + x_{i_m})/2 ,$$

be the midpoints of triangle sides $\overline{x_i x_{i_m}}$.

Using the following quadrature formulas

$$\int_{\partial \mathscr{X}_i} k(x)\frac{\partial u}{\partial \nu_i}\, ds' \approx \sum_{i_m \in \mathscr{S}(i)} l_{ii_m} k_{ii_m} \frac{y_{i_m} - y_i}{h_{ii_m}} , \qquad (4.73)$$

$$\int_{\partial \mathscr{X}_i} u \frac{\partial \varphi}{\partial \nu_i}\, ds' \approx \sum_{i_m \in \mathscr{S}(i)} (\alpha_{ii_m} y_i + (1- \alpha_{ii_m})y_{i_m}) \frac{\varphi_{i_m} - \varphi_i}{h_{ii_m}} , \quad (4.74)$$

where $k_{ii_m} = k(x_{ii_m})$, $\varphi_i = \varphi(x_i)$ and $\alpha_{ii_m} \in [0,1]$. The introduced α_{ii_m} are upwind parameters, which are specified later on. Thus, we derive an FDM approximation to Problem 4.4.

<u>Approximation 4.4.</u> (upwinded difference method over triangulized regions)

$$\left(\sum_{i_m \in \mathscr{S}(i)} \frac{l_{ii_m} k_{ii_m}}{h_{ii_m}} \right) y_i - \sum_{i_m \in \mathscr{S}(i)} \frac{l_{ii_m} k_{ii_m}}{h_{ii_m}} y_{i_m} +$$

$$\left(\sum_{i_m \in \mathscr{S}(i)} \alpha_{ii_m} \frac{\varphi_i - \varphi_{i_m}}{h_{ii_m}} \right) y_i + \sum_{i_m \in \mathscr{S}(i)} (1- \alpha_{ii_m}) \frac{\varphi_i - \varphi_{i_m}}{h_{ii_m}} y_{i_m} = 0,$$

for each $x_i \in \omega_h$, and

$$y_i = g_i ,$$

for each $x_i \in \gamma_h$.

Let $y = (y_1,\ldots,y_{n_1},y_{n_1+1},\ldots,y_n)^T$, then the matrix form of the diffe-rence equation system is a priori of the form (3.18), that is

$$A\,y \;=\; \begin{pmatrix} I & 0 \\ A_{21} & A_{22} \end{pmatrix} y \;=\; f\,, \tag{4.75}$$

where $f = (g_1,\ldots,g_{n_1},0,\ldots,0)^T$.

From the structure of the difference equations it is clear that the part $(A_{21}\ A_{22})$ of A can be split into the sum

$$(A_{21}\ A_{22}) \;=\; (K_{21}\ K_{22}) \;+\; (\Phi_{21}\ \Phi_{22})\,, \tag{4.76}$$

where the first term contains all the expressions with k_{ij} and the second all those with $(\varphi_i - \varphi_j)/h_{ij}$.

For each triangle side $\overline{x_i x_j}$, $x_i, x_j \in \omega_h$, the matrices $(K_{21}\ K_{22})$ and $(\Phi_{21}\ \Phi_{22})$ are built by elementary matrices like that

$$\frac{l_{ij}k_{ij}}{h_{ij}}\begin{pmatrix} 1 & -1 \\ -1 & 1 \end{pmatrix}\begin{matrix} -\,i \\ -\,j \end{matrix} \quad , \quad \frac{\varphi_i - \varphi_j}{h_{ij}}\begin{pmatrix} \alpha_{ij} & (1-\alpha_{ij}) \\ -\alpha_{ij} & -(1-\alpha_{ij}) \end{pmatrix}\begin{matrix} -\,i \\ -\,j \end{matrix} \quad , \tag{4.77}$$
$$\qquad\quad \underset{y_i}{}\ \underset{y_j}{} \qquad\qquad\qquad\qquad \underset{y_i}{}\qquad \underset{y_j}{}$$

respectively.

Hence, we have

$$A \;=\; K + \Phi \;=\; \begin{pmatrix} I & 0 \\ K_{21} & K_{22} \end{pmatrix} + \begin{pmatrix} 0 & 0 \\ \Phi_{21} & \Phi_{22} \end{pmatrix}.$$

In the following we shall analyse properties of the matrix $A = K + \Phi$. First, it follows from (4.77) that $K_{21} \leqslant 0$ and that $K_{21}^T e'' < 0'$, where e'' is the $(n-n_1)$-dimensional vector of all ones and $0'$ is the n_1-dimensional vector of all zeros. Further, we have $K_{22} = K_{22}^T$ which is an irreducible weakly diagonally dominant L-matrix. Thus, K_{22} is a nonsingular M-matrix and such is K. Additionally, we see that K is weakly row diagonally dominant in the following manner

$$K\,e \;=\; (1,\ldots,1,0,\ldots,0)^T \geqslant 0.$$

Second, suppose that $\Phi \neq 0$. Then it is immediately seen from (4.77) that $\Phi \neq \Phi^T$ and that $\Phi^T e = 0$ for each choice of the upwind parameters $\alpha_{ij} \in [0,1]$. Now we choose α_{ij} such that $\Phi \in \mathscr{L}_c^{n \times n}$, see Section 3.2.

It is readily verified that $\Phi \in \mathscr{L}_c^{n \times n}$ if α_{ij} are given by

$$\alpha_{ij} \;=\; \begin{cases} 1 & \text{if } \varphi_i - \varphi_j > 0\,, \\ 0 & \text{if } \varphi_i - \varphi_j < 0\,. \end{cases} \tag{4.78}$$

Thus, by virtue of Theorem 3.13., $A = K + \Phi$ is a nonsingular M-matrix.

We next discuss some properties of the difference equation system $Ay = f$ under the assumption that $A = K + \Phi$ is the nonsingular M-matrix from above.

Let $f \geqslant 0$, i.e. $g(x) \geqslant 0$ for $x \in \partial\Omega$. Then the solution y is nonnegative over $\overline{\omega}_h$, that is $y = A^{-1}f \geqslant 0$.

Next suppose that the potential φ is constant, i.e. $\varphi(x) = \text{const}$ for $x \in \Omega$. This implies $\Phi = 0$ and $Ay = Ky = f$ satisfies the discrete boundary maximum principle by Theorem 3.21. That is, for $y = K^{-1}f$,

$$\min_{x_k \in \gamma_h} g_k \quad \leqslant \quad y_i \quad \leqslant \quad \max_{x_k \in \gamma_h} g_k \qquad \text{for each } x_i \in \overline{\omega}_h. \qquad (4.79)$$

Let $k(x) \equiv 1$ for $x \in \Omega$, then $y = K^{-1}f$ approximates a harmonic function, for which (4.60) holds.

It is also relevant to note that there is a more general possibility for choosing $\alpha_{ij} \in [0,1]$ such that $A = K + \Phi$ is a nonsingular weakly column diagonally dominant M-matrix. In so doing, we suppose all off-diagonal entries of A nonpositive, that is, we deduce the following inequalities according to α_{ij} from (4.77),

$$- l_{ij}k_{ij} - \alpha_{ij}(\varphi_i - \varphi_j) \leqslant 0 \qquad \text{if} \quad \varphi_i - \varphi_j < 0,$$

or

$$- l_{ij}k_{ij} + (1 - \alpha_{ij})(\varphi_i - \varphi_j) \leqslant 0 \qquad \text{if} \quad \varphi_i - \varphi_j > 0,$$

$$(4.80)$$

for each triangle side $\overline{x_i x_j}$ with $x_i, x_j \in \omega_h$. It is then quite easy to see that (4.80) includes the special choice (4.78). However, we do not pay them further attention.

We shall now briefly indicate how to proceed in the FDM approximation of the following two-dimensional problems.

Problem 4.5.

$$\tilde{L}u = Lu + q(x)u = \tilde{f}(x), \qquad x \in \Omega,$$
$$u = g(x), \qquad x \in \partial\Omega.$$

Suppose that L is given by one of the Problems 4.3. or 4.4. and that the smoothness of $q(x) \geqslant 0$, $\tilde{f}(x)$ for $x \in \Omega$ guarantees a unique solution $u(x) \in C^2(\Omega) \cap C(\overline{\Omega})$ of Problem 4.5.

First, assume that L comes from Problem 4.3. In this case, Problem 4.5. satisfies the following maximum principle, see [24]. Letting $\tilde{f}(x) \leqslant 0$ $(\tilde{f}(x) \geqslant 0)$ for $x \in \Omega$, a nonnegative maximum (a nonpositive

minimum) of $u(x)$, if it exists, is taken on at the boundary $\partial\Omega$.
Return now to the Approximation 4.3. According to Problem 4.5. it
changes in the following manner. For $x_{ij} \in \omega_h$, we have to add the
term $q_{ij}y_{ij}$ on the left-hand side of the difference equations, $q_{ij} \geq 0$
by the assumptions, and the right-hand side takes the form $\tilde{f}_{ij} = \tilde{f}(x_{ij})$. Thus, from the difference equation system $Ay = f$ to Problem
4.3. we derive a new difference equation system $\tilde{A}y = \tilde{f}$, putting
$\tilde{A} = A + Q$, $Q = \text{diag}(.,0,...,q_{ij},...) \geq 0$ and $\tilde{f}$ is composed of the
components g_{ij} and $\tilde{f}_{ij}$.
By virtue of Theorem 3.12., the matrix $\tilde{A} = A + Q$ then is also a nonsingular M-matrix and $\tilde{A}^{-1} \leq A^{-1}$ holds. For $\tilde{A}y = \tilde{f}$ an inequality of
the type of inequality (4.54) is deducible. The proof is virtually
the same as the one we have given to show that (4.54) holds in the
one-dimensional case.

Second, let L be given by Problem 4.4. Based on the Approximation
4.4. over a triangulation of $\bar{\Omega}$ of weakly acute type, see also
Figure 4.9.,we now construct an FDM approximation to Problem 4.5.
We see that the integration of $\tilde{L}u = \tilde{f}(x)$ over $\mathcal{K}_i$ produces two
additive terms in (4.70), (4.72), one on the left- and one on the
right-hand side, which are of the form

$$\int_{\mathcal{K}_i} q(x)\, u\, dx \qquad \text{and} \qquad \int_{\mathcal{K}_i} \tilde{f}(x)\, dx \ ,$$

respectively.
To derive difference equations, we apply Approximation 4.4. and
additionally use the simple quadrature formulas

$$\int_{\mathcal{K}_i} q(x)\, u\, dx \approx \hbar_i q_i y_i \qquad \text{and} \qquad \int_{\mathcal{K}_i} \tilde{f}(x)\, dx \approx \hbar_i \tilde{f}_i \ , \qquad (4.81)$$

with $q_i = q(x_i)$, $\tilde{f}_i = \tilde{f}(x_i)$, $\hbar_i = \text{mes } \mathcal{K}_i > 0$ for each $x_i \in \omega_h$.
Thus, the difference equation system $Ay = f$, defined by (4.75),
changes into the system $\tilde{A}y = \tilde{f}$ putting $\tilde{A} = A + Q$,

$$Q = \text{diag}(0,...,0,\hbar_{n_1+1}q_{n_1+1},...,\hbar_n q_n) \geq 0 \ ,$$

and

$$f = (g_1,...,g_{n_1},\hbar_{n_1+1}f_{n_1+1},...,\hbar_n f_n)^T .$$

By Theorem 3.12., the matrix $\tilde{A} = A + Q$ is a nonsingular M-matrix,
which is weakly column diagonally dominant by (4.77).
Thus, $\tilde{f} \geq 0$ implies $y = \tilde{A}^{-1}f \geq 0$.

It may be noted that, if $q_i > 0$ for $x_i \in \omega_h$ are sufficiently large

such that $\check{A}$ is strongly or weakly row diagonally dominant, we can
directly apply the region boundary maximum principle from Section
3.4.2. We remark that the nonsingular M-matrix A becomes strongly
row diagonally dominant by a positive diagonal matrix D, that is,
$\check{A}De > 0$, which leads to the system $\check{A}Dy' = \tilde{f}$, $y' = D^{-1}y$. For this
we refer to Property 3.4. But the problem is to find explicitly such
a positive diagonal matrix D.

4.3.3. Difference approximations to parabolic problems

Parabolic problems arise in irreversible time-dependent processes
and are defined, therefore, in space-time cylinders $\Omega \times (0,T) \subset R^d \times R^1_+$.
Suppose here that $\Omega \times (0,T)$ is bounded, that is, $\Omega \subset R^d$ is a bounded
region with $\partial\Omega$ sufficiently smooth for $d \geqslant 2$ and $0 < T < \infty$.
Solutions $u(x,t)$, $(x,t) \in \overline{\Omega \times (0,T)}$, of parabolic initial boundary
value problems obey a maximum principle. A detailed description may
be found in [13,22,24] and elsewhere.
The parabolic equations themselves represent the differential form
of conservation laws. For example, we refer to conservation of mass
in diffusion processes and to conservation of heat in heat conduc-
tion processes.

Our object in the present section is to consider FDM approximations
to parabolic problems stated by (4.5). For this, we shall assume an
FDM space grid $\overline{\omega}_h = \omega_h + \gamma_h \subset \overline{\Omega}$ and a discretization of the
time interval $[0,T]$ by a time grid $\overline{\omega}_\tau$, which is defined by

$$\overline{\omega}_\tau = \left\{ t_j : \; 0 = t_0 < t_1 < \ldots < t_{m-1} < t_m = T \right\}.$$

Thus, $\overline{\Omega \times (0,T)}$ is discretized by the grid $\overline{\omega}_h \times \overline{\omega}_\tau$.

In the following, we shall only use implicit difference schemes, i.e.
schemes with backward finite difference approximation to the time
derivative $\frac{\partial u}{\partial t}$.

Therefore, it suffices to describe how the new approximation $y(t_{j+1})$
to $u(x,t_{j+1})$ over $\overline{\omega}_h$ is then computed from the previous approximation $y(t_j)$, $t_{j+1} = t_j + \tau_j$.

For convenience, we simplify the notation. Henceforth, let

$$t = t_{j+1}, \quad \check{t} = t_j, \quad \tau = \tau_j, \quad t = \check{t} + \tau ,$$

and

$$y = y(\check{t} + \tau) = (y_i)_{x_i \in \overline{\omega}_h} , \quad \check{y} = y(\check{t}) = (\check{y}_i)_{x_i \in \overline{\omega}_h} , \quad (4.82)$$

We start our considerations with some general remarks.
Let

$$\frac{\partial u}{\partial t} + L\,u = \tilde{f}(x,t)\ , \qquad (x,t) \in \Omega \times (0,T)$$

be a parabolic differential equation. Then, we consider the approximation

$$\frac{y_i - \check{y}_i}{\tau} + (L_h y)_i = \tilde{f}_i\ , \qquad x_i \in \omega_h, \qquad (4.83)$$

putting $\tilde{f}_i = \tilde{f}(x_i, \check{t} + \tau)$.

We assume that the FDM approximation $(L_h y)_i = \tilde{f}_i$, $x_i \in \omega_h$ of the elliptic part leads to a difference equation system $A y = \tilde{f}$ with a nonsingular M-matrix A, the approximation of the Dirichlet boundary conditions of the problem incorporated.
Then, from (4.83) we have

$$\left(\frac{1}{\tau} I' + A\right) y = \tilde{f} + \frac{1}{\tau}\check{y}\ , \qquad (4.84)$$

the difference equation systems for each $t = t_j$, $j=1,\dots,m$.
The matrix I' is a diagonal matrix with diagonal entries 1 if $x_i \in \omega_h$ and 0 if $x_i \in \gamma_h$. In analogy, $\check{y}$ is a vector with components $\check{y}_i$ if $x_i \in \omega_h$ and 0 if $x_i \in \gamma_h$.
Thus, the system matrix $\left(\frac{1}{\tau} I' + A\right)$ of (4.84) is a nonsingular M-matrix by Theorem 3.12.
Additionally, let A be a weakly row diagonally dominant M-matrix. Then $\left(\frac{1}{\tau} I' + A\right)$ is in many cases a strongly row diagonally dominant M-matrix such that it is possible to apply the region maximum principle from Section 3.4.2. to deduce upper and lower solution bounds to y.
Our next aim is to illustrate this approach by a few samples.
We begin with a one-dimensional example. For definiteness, suppose $\Omega = (0,1)$ with $\partial\Omega = \{0,1\}$.
We let

$$\partial\Omega_T = (\Omega \times \{t = 0\}) \cup (\partial\Omega \times (0,T])\ ,$$

the parabolic boundary of the space-time cylinder $\Omega \times (0,T)$.

<u>Problem 4.6.</u>

$$\begin{aligned}
\frac{\partial u}{\partial t} + L\,u &= 0\ , & \Omega \times (0,T]\ , \\
u(x,0) &= u_0(x)\ , & \Omega\ , \\
u(0,t) &= g_0(t)\ , & 0 < t \leq T\ , \\
u(1,t) &= g_1(t)\ , & 0 < t \leq T\ .
\end{aligned}$$

It is assumed here that L is given by Problem 4.1. with $b = b(x,t)$
and that Problem 4.6. satisfies the compatibility conditions

$$u_0(0) = g_0(0) , \quad u_0(1) = g_1(0) .$$

Furthermore, the assumption is made that Problem 4.6. admits a
classical solution $u(x,t)$.

It is a well-known fact that in the case $b \equiv 0$ the maximum principle
takes the following form. Both $\max u(x,t)$ and $\min u(x,t)$ in
$\overline{\Omega} \times (0,T)$ are taken on at points in the parabolic boundary $\partial \Omega_T$,
see [22].

Now suppose that $\overline{\Omega}$ is discretized by a uniform grid $\overline{\omega}_h = \omega_h + \gamma_h$
with step size $h = 1/n$. To approximate the elliptic part Lu of the
parabolic differential equation over ω_h, we apply, for instance,
the artificial diffusion scheme given by Approximation 4.1.c.
We thus arrive at a nonsingular tridiagonal M-matrix A, which is
entirely defined in Approximation 4.1.c and for which

$$A e = (1,0,\dots,0,1)^T \geq 0$$

holds.

To generate the system (4.84), let

$$I' = \operatorname{diag}(0,1,\dots,1,0) , \quad \check{y} = (0,\check{y}_1,\dots,\check{y}_{n-1},0)^T,$$

$$\tilde{f} = (g_0(t),0,\dots,0,g_1(t))^T.$$

Hence, we have

<u>Approximation 4.6.</u>

$$(I' + \tau A) y = \check{y} + \tau \tilde{f} . \qquad (4.85)$$

It is then quite easy to see that the system matrix $(I' + \tau A)$
is a strongly row diagonally dominant M-matrix, i.e.

$$(I' + \tau A) e = (\tau,1,\dots,1,\tau)^T > 0.$$

Let $B = \operatorname{diag}(\tau,1,\dots,1,\tau)$, we rewrite (4.85) as

$$(I' + \tau A) y = B v ,$$

putting $v = B^{-1}(\check{y} + \tau \tilde{f}) = (g_0(t),\check{y}_1,\dots,\check{y}_{n-1},g_1(t))^T.$

Applying Theorem 3.24. gives

$$\min\left\{g_0(t),g_1(t), \min_{1 \leq j \leq n-1} \check{y}_j\right\} \leq y_i \leq \max\left\{g_0(t),g_1(t), \max_{1 \leq j \leq n-1} \check{y}_j\right\}, \quad (4.86)$$

for $i=0,\dots,n$. Thus, at each time step the discrete maximum principle
(4.86) holds for Approximation 4.6.

Next, we briefly consider Problem 4.6. with changed boundary conditions. For this, we state at $\partial\Omega \times (0,T]$

$$\alpha_0 u(0,t) - \beta_0 \frac{\partial u(0,t)}{\partial x} = g_0(t) \ , \quad \alpha_1 u(1,t) + \beta_1 \frac{\partial u(1,t)}{\partial x} = g_1(t), \quad (4.87)$$

where we assume $\alpha_1 > 0$, $\beta_1 \geqslant 0$, $i=1,2$.

Then, the following simple difference approximation to (4.87)

$$(\alpha_0 + \frac{\beta_0}{h}) \ y_0 \ - \ \frac{\beta_0}{h} \ y_1 \ = \ g_0(t) \ ,$$

$$(4.88)$$

$$- \frac{\beta_1}{h} \ y_{n-1} + (\alpha_1 + \frac{\beta_1}{h}) \ y_n \ = \ g_1(t) \ ,$$

is incorporated into Approximation 4.6. instead of the equations $y_0 = g_0(t)$ and $y_n = g_1(t)$. Let us denote by A' the new tridiagonal M-matrix, for which

$$\tau A' e \ = \ (\tau \alpha_0, 0, \ldots, 0, \tau \alpha_1)^T \geqslant 0$$

holds.

To this end, we have

$$(I' + \tau A') \ y \ = \ \breve{y} \ + \ \tau \tilde{f} \ = \ B \, v, \qquad (4.89)$$

putting $B = \mathrm{diag}(\tau \alpha_0, 1, \ldots, 1, \tau \alpha_1) \geqslant 0$, $v = B^{-1}(\breve{y} + \tau \tilde{f}) = (g_0(t)/\alpha_0, \breve{y}_1, \ldots, \breve{y}_{n-1}, g_1(t)/\alpha_1)^T$.

It has been found that for the difference equation system (4.89) the conditions of Theorem 3.24. hold, letting $w = 0$. Thus, the solution $y = (y_0, \ldots, y_n)^T$ of (4.89) is bounded as follows

$$\min\left\{ \frac{g_0(t)}{\alpha_0}, \frac{g_1(t)}{\alpha_1}, \min_{1 \leqslant j \leqslant n-1} \breve{y}_j \right\} \leqslant y_i \leqslant \max\left\{ \frac{g_0(t)}{\alpha_0}, \frac{g_1(t)}{\alpha_1}, \max_{1 \leqslant j \leqslant n-1} \breve{y}_j \right\}, \quad (4.90)$$

for each $i=0,\ldots,n$.

We are now ready to consider a more general parabolic problem and its FDM discretization.

Problem 4.6.

$$\frac{\partial u}{\partial t} \ + \ L u \ + \ q(x,t) \, u \ = \ \tilde{f}(x,t) \ , \qquad \Omega \times (0,T) \ ,$$

$$u(x,0) \ = \ u_0(x) \ , \qquad \Omega \ ,$$

$$\alpha_s u(s,t) + (-1)^{s+1} \beta_s \frac{\partial u(s,t)}{\partial x} \ = \ g_s(t) \ , \qquad 0 < t \leqslant T, \quad s=0,1.$$

Let L be the same as in Problem 4.6., i.e. L is given by Problem 4.1.

Further, assume $q(x,t) \geq 0$, $\tilde{f}(x,t)$, $g_s(t)$, $s=0,1$ sufficiently smooth, $\alpha_s > 0$, $\beta_s \geq 0$, $s=0,1$ such that Problem 4.6.´ has a classical solution $u(x,t)$.

We deduce an FDM approximation to Problem 4.6.´ based on Approximation 4.6. with reference to (4.88).

Let

$$Q = \mathrm{diag}(0,q_1,\ldots,q_{n-1},0) \geq 0, \qquad q_i = q(x_i,t),$$

$$\tilde{\tilde{f}} = (0,\tilde{f}_1,\ldots,\tilde{f}_{n-1},0)^T, \qquad \tilde{f}_i = \tilde{f}(x_i,t),$$

the finite difference equation system can then be written in the form

$$(I´ + \tau A´ + \tau Q) y = \breve{y} + \tau \tilde{f} + \tau \tilde{\tilde{f}}. \tag{4.91}$$

By virtue of Theorem 3.12., $(I´ + \tau A´ + \tau Q)$ is a nonsingular M-matrix and

$$0 \leq (I´ + \tau A´ + \tau Q)^{-1} \leq (I´ + \tau A´)^{-1}$$

holds.

Thus, the system (4.91) has a unique solution $y \geq 0$ if the right-hand side vector is nonnegative.

By the assumption $q(x,t) \geq 0$, we have

$$(I´ + \tau A´ + \tau Q) e = (\tau\alpha_0, 1+\tau q_1,\ldots,1+\tau q_{n-1}, \tau\alpha_1)^T > 0.$$

Now, let $B = C = \mathrm{diag}(\tau\alpha_0, 1+\tau q_1,\ldots,1+\tau q_{n-1}, \tau\alpha_1) \geq 0$, we re-write the system (4.91) as follows

$$(I´ + \tau A´ + \tau Q) y = B v + \tau C w, \tag{4.92}$$

where

$$v = B^{-1}(\breve{y} + \tau f) = \left(\frac{g_0(t)}{\alpha_0}, \frac{\breve{y}_1}{1+\tau q_1}, \cdots, \frac{\breve{y}_{n-1}}{1+\tau q_{n-1}}, \frac{g_1(t)}{\alpha_1} \right)^T,$$

$$w = C^{-1}\tilde{\tilde{f}} = \left(0, \frac{\tilde{f}_1}{1+\tau q_1}, \cdots, \frac{\tilde{f}_{n-1}}{1+\tau q_{n-1}}, 0 \right)^T.$$

Hence, by Theorem 3.24., we find

$$\min\left\{ \frac{g_0(t)}{\alpha_0}, \frac{g_1(t)}{\alpha_1}, \min_{1 \leq j \leq n-1} \frac{\breve{y}_j}{1+\tau q_j} \right\} + \tau \min\left\{ 0, \min_{1 \leq j \leq n-1} \frac{\tilde{f}_j}{1+\tau q_j} \right\} \leq y_i \leq$$

$$\max\left\{ \frac{g_0(t)}{\alpha_0}, \frac{g_1(t)}{\alpha_1}, \max_{1 \leq j \leq n-1} \frac{\breve{y}_j}{1+\tau q_j} \right\} + \tau \max\left\{ 0, \max_{1 \leq j \leq n-1} \frac{\tilde{f}_j}{1+\tau q_j} \right\},$$

for each $i=0,\ldots,n$.

The above results may be summarized as follows. If the implicit
difference method (4.83) directly leads to difference equation systems which involve strongly row diagonally dominant M-matrices, we
can immediately deduce lower and upper solution bounds at each time
step by means of the region maximum principle. On the other hand, if
the system matrices are not strongly row diagonally dominant, we have
to look for a positive diagonal matrix D which realizes the latent
strongly row diagonal dominance, see Property 3.5. For instance, this
is the case in Problem 4.6.´, assuming $\alpha_s = 0$, $\beta_s > 0$, $s=0,1$.

Parabolic initial boundary value problems involve conservation laws,
see [13,24]. Here, we state a suitable formulation of them and show
how FDM approximations via nonsingular M-matrices represent discrete
versions of the conservation laws.
Let $\Omega \subset R^d$ be a bounded region with boundary $\partial\Omega$ sufficiently smooth.
For smooth $k(x,t) \geqslant k_o = \text{const} > 0$ and potential $\varphi(x,t)$ let

$$\vec{w} = - k(x,t) \nabla u + \nabla\varphi(x,t) u .$$

Physically speaking, $\vec{w}$ may be a mass or heat flux in the space-time
cylinder $\Omega \times (0,T)$.
Then, the conservation law is given by the parabolic differential
equation

$$\frac{\partial u}{\partial t} + \nabla\vec{w} = 0 , \qquad \Omega \times (0,T) , \qquad (4.93)$$

omitting possible source terms.
Now consider a time interval $[t_1,t_2] \subset [0,T]$ and an arbitrary sub-
region Ω' of Ω with $\partial\Omega'$ piecewise smooth. Further, denote by ν
the outward unit normal to $\partial\Omega' \times (t_1,t_2)$.
By integration of the parabolic differential equation (4.93) over
$\Omega' \times (t_1,t_2)$ we get

$$\int_{\Omega'} u(x,t_2)\, dx = \int_{\Omega'} u(x,t_1)\, dx + \int_{\partial\Omega' \times (t_1,t_2)} \frac{\partial\vec{w}}{\partial\nu}\, ds, (4.94)$$

which may be interpreted as the integral form of the conservation
law under consideration.
With these preparations, let us now turn our attention to the
following parabolic problem.
Problem 4.7.

$$\frac{\partial u}{\partial t} + \nabla\vec{w} = 0, \qquad \Omega \times (0,T] ,$$
$$u(x,0) = u_o(x), \qquad \Omega ,$$
$$\frac{\partial\vec{w}}{\partial\nu} = g(x,t) , \qquad \partial\Omega \times (0,T] .$$

According to Problem 4.7., we get from (4.94)

$$\int_{\Omega} u(x,t_2)\, dx \;=\; \int_{\Omega} u(x,t_1)\, dx \;+\; \int_{\partial\Omega\times(t_1,t_2)} g(x,t)\, ds \; . \qquad (4.95)$$

Let $g(x,t) \equiv 0$ on $\partial\Omega\times(t_1,t_2)$, then the boundary condition $\dfrac{\partial u}{\partial\nu} = 0$ is to be interpreted as "no flux" condition. Thus

$$\int_{\Omega} u(x,t_2)\, dx \;=\; \int_{\Omega} u(x,t_1)\, dx \qquad \text{for each } 0 \le t_1 < t_2 \le T \; .$$

Next, we construct an FDM approximation to Problem 4.7., which very obviously involves a discrete version of the conservation law.

For this, we assume $\Omega \subset R^2$ as a convex polygonal region and employ Approximation 4.4. together with all the notations introduced.
As in Approximation 4.4., let $\overline{\omega}_h = \omega_h + \gamma_h$ be the set of all vertices x_i of the triangles generated by triangulation of $\overline{\Omega}$ of weakly acute type. According to Figure 4.9., elementary regions $\mathcal{H}_i$ are defined for each $x_i \in \omega_h$, and, additionally, are also defined for $x_i \in \gamma_h$, see Figure 4.10.

Fig.4.10.

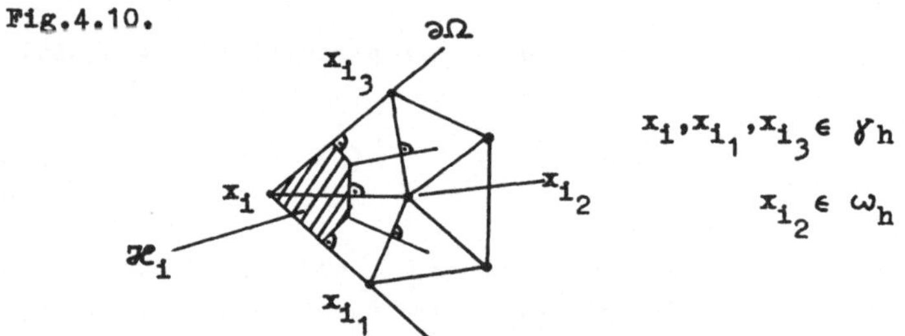

Thus $\displaystyle\sum_{x_i \in \overline{\omega}_h} \hbar_i \;=\; \text{mes } \Omega \quad , \quad \hbar_i \;=\; \text{mes } \mathcal{H}_i \; .$

For each $x_i \in \gamma_h$, we split the difference star $\mathcal{S}(i)$ into

$$\mathcal{S}(i) \;=\; \mathcal{S}_{\omega_h}(i) \;+\; \mathcal{S}_{\gamma_h}(i) \; ,$$

where $\mathcal{S}_{\omega_h}(i) = \mathcal{S}(i) \cap \omega_h$ and $\mathcal{S}_{\gamma_h}(i) = \mathcal{S}(i) \cap \gamma_h$.

We observe

$$\mathcal{S}_{\gamma_h}(i) \neq \emptyset \qquad \text{for each } x_i \in \gamma_h \quad \text{and assume} \quad \mathcal{S}_{\omega_h}(i) \neq \emptyset \; .$$

Now we are ready to derive difference equations from (4.94) via the
following quadrature formulas, putting $\Omega' = \mathcal{H}_i$ for each $x_i \in \overline{\omega}_h$
and $t_1 = \check{t}$, $t_2 = \check{t} + \tau$. Let

$$\int_{\mathcal{H}_i} u(x,\check{t}+\tau)dx \approx \hbar_i y_i \ , \qquad \int_{\mathcal{H}_i} u(x,\check{t})dx \approx \hbar_i \check{y}_i \qquad \text{for each } x_i \in \overline{\omega}_h,$$

$$\int_{\partial\mathcal{H}_i \times (\check{t},\check{t}+\tau)} \frac{\partial \omega}{\partial \nu} ds \approx \tau\cdot(\text{upwinded difference method of Approximation 4.4.})$$

$$\text{for each } x_i \in \omega_h,$$

$$\int_{\partial\mathcal{H}_i \times (\check{t},\check{t}+\tau)} \frac{\partial \omega}{\partial \nu} ds \approx \tau\cdot(\text{upwinded difference method of Approximation 4.4.})$$
$$\text{for } i_m \in \mathcal{S}_{\omega_h}(i)$$
$$+ \ \tau g_i s_i \ , \qquad\qquad \text{for each } x_i \in \gamma_h,$$

where $g_i = g(x_i,\check{t}+\tau)$, $s_i = \displaystyle\sum_{i_m \in \mathcal{S}_{\gamma_h}(i)} h_{ii_m}/2$.

We remark that $\displaystyle\sum_{x_i \in \gamma_h} s_i = \text{mes } \partial\Omega$.

For brevity, we shall then put all the arising difference equations
into matrix form immediately.

Let $y = (y_1,\ldots,y_{n_1},y_{n_1+1},\ldots,y_n)^T$ be the vector of the approximate
values at time level given by $\check{t}+\tau$, i.e. $y_i \approx u(x_i,\check{t}+\tau)$, $x_i \in \gamma_h$
for $i=1,\ldots,n_1$ and $x_i \in \omega_h$ for $i=n_1+1,\ldots,n$.
Further, let

$$H = \text{diag}(\hbar_1,\ldots,\hbar_{n_1},\hbar_{n_1+1},\ldots,\hbar_n) \ , \qquad\qquad (4.96)$$

which is a positive diagonal matrix and thus a nonsingular M-matrix.
Then, the whole difference equation system is given by

$$A y = (H + \tau K + \tau \Phi) y = H \check{y} + \tau g \ . \quad (4.97)$$

The matrices K and Φ are generated by (4.77), $\check{y} = (\check{y}_1,\ldots,\check{y}_n)^T$
contains the previous approximate values at time level $\check{t}$, i.e.
$\check{y}_i \approx u(x_i,\check{t})$ and the components of the vector g, where

$$g = (s_1 g_1,\ldots,s_{n_1} g_{n_1},0,\ldots,0)^T,$$

bring in the boundary condition influence.
Next, the properties of the system matrix $A = H + \tau K + \tau \Phi$ will be
properly assessed.

First, we note that $K = K^T$ and if $\tilde{\Phi} \neq 0$, then $\tilde{\Phi} \neq \tilde{\Phi}^T$. Further, we can see from (4.77) that $Ke = K^Te = 0$ and $\tilde{\Phi}^Te = 0$ for any choice of the upwind parameters $\alpha_{ij} \in [0,1]$.

Now adding up all the difference equations of the system (4.97), we get

$$\sum_{x_i \in \bar{\omega}_h} \hbar_i y_i \quad = \quad \sum_{x_i \in \bar{\omega}_h} \hbar_i \check{y}_i \quad + \quad \tau \sum_{x_i \in \gamma_h} s_i g_i \ , \qquad (4.98)$$

which may be interpreted as a discrete version of the conservation law (4.95).

After this, we shall investigate the conditions which guarantee that $A = H + \tau K + \tau \tilde{\Phi}$ is a nonsingular M-matrix.

From the above, we have $\tau K \in \mathcal{L}_r^{n \times n}$. By virtue of Theorem 3.13., it follows that $H + \tau K$ is a Stieltjes matrix because H is a strongly row diagonally dominant M-matrix.

Suppose that $\tilde{\Phi} \neq 0$. Let the upwind parameters α_{ij} be chosen by (4.78) for each triangle side $\overline{x_i x_j}$ with at least one vertex in ω_h, then $\tau \tilde{\Phi} \in \mathcal{L}_c^{n \times n}$, and by the remark to Theorem 3.13., we derive that $A = H + \tau K + \tau \tilde{\Phi}$ is a nonsingular M-matrix because of $(H + \tau K)^Te = He > 0$.

In summary, $H + \tau K$ is a Stieltjes matrix and $A = H + \tau K + \tau \tilde{\Phi}$ is a strongly column diagonally dominant M-matrix if $\tilde{\Phi} \neq 0$.

The more general choice (4.80) of α_{ij} does not change the above mentioned properties of A.

We note that for $\tilde{\Phi} = 0$ we can immediately deduce lower and upper solution bounds in each time step, applying the region maximum principle.

4.4. Finite element methods

In our discussion concerning nonsingular M-matrices in FEM approximations, we shall inquire into some problems which are quite similar to those of Section 4.3. But here they are stated a priori in a weak formulation.

According to elliptic problems (4.3), let the weak form for the moment be described as an abstract variational problem. That is, loosely speaking, we search for a unique function $u \in U$, the weak solution of problem (4.3), such that

$$a(u,v) \quad = \quad l(v) \qquad \text{for each } v \in V \ , \qquad (4.99)$$

where U,V are suitable function spaces, $a(u,v)$ is a bilinear and $l(v)$ is a linear functional on U,V. The relation (4.99) follows from the elliptic differential equation $Lu = f$ multiplied by some

$v \in V$ and then integrated by parts. Quite clearly, such properties of
$a(u,v)$ as symmetry, boundedness or definiteness mainly depend on the
properties of L. Of the very extensive literature dealing with this
approach we only cite [4,11,25].

On the basis of the waek formulation of the elliptic problems under
consideration, the application of the FEM begins with the subdivision
of the region Ω into smaller pieces $\mathcal{E}_i$ called finite elements.
In the one-dimensional case, $\Omega = (a,b)$, the finite elements $\mathcal{E}_i$ are
subintervals. Two-dimensional regions Ω are usually divided into
triangular or quadrilateral finite elements $\mathcal{E}_i$, with adjacent elements sharing a common side. Further, we have to define a set of nodal points $\{x_i\}_{i \in N}$, which is, in connection with the finite element
division, the basis for the approximate solution.

Constructing a polynomial for each finite element $\mathcal{E}_i$, we derive a
piecewise polynomial defined on the finite element division of Ω ,
and since adjacent elements have common nodal points, it will have a
certain degree of continuity automatically imposed across interele-
ment boundaries.

It is easy to verify that this procedure amounts to constructing a
set of basic functions $\varphi_i(x)$, one for each nodal point x_i, $i \in N$,
whose support is limited to finite elements sharing the corresponding node x_i. The piecewise polynomial can thus be expressed in the
form of a linear combination of the basic functions

$$u_h(x) \;=\; \sum_{i \in N} y_i \, \varphi_i(x) \, . \tag{4.100}$$

Let $U_h = \mathrm{span}\left\{\varphi_i(x)\right\}_{i \in N} \subset U$ be some finite dimensional subspace,
called trial-space. Further, we introduce a finite dimensional test-
space $V_h \subset V$, such that (4.99) can be approximated as follows.

We search for a unique element $u_h(x) \in U_h$ such that

$$a(u_h, \psi) \;=\; l(\psi) \quad \text{for each} \quad \psi \in V_h \tag{4.101}$$

holds.

This FEM approach is usually called the Galerkin method if U_h and V_h
are generated by the same type of basic functions $\varphi_i(x)$, and the
Galerkin-Petrov method if the test-functions $\psi_i(x)$, $V_h = \mathrm{span}\left\{\psi_i(x)\right\}$
differ from the trial-functions $\varphi_i(x)$.

For the convergence properties, we refer the reader to [4,11,25,37]
and elsewhere.

Other types of FEM approximations will not be taken into account
here, see [4]. We note that weak formulations of parabolic problems
will be considered below.

4.4.1. Finite element approximations to one-dimensional elliptic boundary value problems

Let $\Omega = (0,1)$ be subdivided into finite elements $\mathcal{E}_i = (x_{i-1}, x_i)$, $i \in N$. For sake of simplicity, assume $x_i - x_{i-1} = h = 1/n$, $\forall\, i \in N$ with nodal points $x_i = ih$, $i=0,\ldots,n$.
We start with a weak formulation of Problem 4.1.

Problem 4.8.

Search for $u \in U = \left\{ u \in W_2^1(0,1): \ u(0) = u_0, \ u(1) = u_1 \right\}$ such that

$$a(u,v) \ = \ \int_0^1 (\, u'v' + b(x)u'v \,) \, dx \ = \ 0, \quad \forall\, v(x) \in V,$$

$V = W_2^1(0,1)$.

It is readily seen that $a(u,v)$ is nonsymmetric if $b(x) \neq 0$, which is then also a main property of the derived FEM system.
First, we apply the Galerkin method to approximate Problem 4.8. For this, we define the following basic functions

$$\varphi_i(x) \ = \ \begin{cases} (x - x_{i-1})/h \ , & x \in \mathcal{E}_i \ , \\ (x_{i+1} - x)/h \ , & x \in \mathcal{E}_{i+1} \ , \\ 0 \ \text{ otherwise} \ , \end{cases} \qquad (4.102)$$

for $i=1,\ldots,n-1$ and $\varphi_0(x)$, $\varphi_n(x)$ on $\mathcal{E}_1$ and $\mathcal{E}_n$, respectively.
Thus

$$U_h \ = \ \left\{ u_h(x) = \sum_{i=0}^{n} y_i\, \varphi_i(x) \ , \quad y_0 = u_0, \ y_n = u_1 \right\} \subset U \ ,$$

$$V_h \ = \ \text{span}\left\{ \varphi_i(x) \right\}_{i=1,\ldots,n-1} \subset V \ .$$

Approximation 4.8.

Search for $u_h(x) \in U_h$ such that

$$a(u_h, \varphi_j) \ = \ 0 \ , \qquad j=1,\ldots,n-1$$

holds.

We obtain the following FEM system

$$y_0 \ = \ u_0 \ ,$$

$$\sum_{i=0}^{n} y_i\, a(\varphi_i, \varphi_j) \ = \ 0 \ , \qquad j=1,\ldots,n-1 \ , \qquad (4.103)$$

$$y_n \ = \ u_1 \ ,$$

and derive its matrix form $Ay = f$, putting $y = (y_0,\ldots,y_n)^T$, $A=(a_{ji})$, $f = u_0 e_1 + u_1 e_{n+1}$. The entries of the tridiagonal matrix A are thus defined by

$$a_{ji} = a(\varphi_i, \varphi_j) = \int_0^1 (\varphi_i' \varphi_j' + b(x)\,\varphi_i'\,\varphi_j)\,dx , \qquad (4.104)$$

for $i=0,..,n$, $j=1,..,n-1$.

Some interesting properties of A can be easily illustrated by considering the simplest case, namely, $b(x) = b = $ const. Thus, we readily compute

$$a_{ji} = \begin{cases} -\,1/h - b/2 , & i=j-1, \\ 2/h , & i=j, \\ -\,1/h + b/2 , & i=j+1, \\ 0 , & |i - j| > 1. \end{cases} \qquad (4.105)$$

In this case, and also for

$$\int_0^1 b(x)\,\varphi_i'\,\varphi_j\,dx \approx b_i \int_0^1 \varphi_i'\,\varphi_j\,dx ,$$

with $b_i = b(x_i)$, the resulting FEM system $Ay = f$ is completely equivalent to Approximation 4.1.a. Thus the system matrix A is a nonsingular M-matrix if and only if condition (4.20)·holds, and we have to impose (4.21) to guarantee the boundary maximum principle (4.22), see also Section 3.4.1.
In summary, Approximation 4.8. suffers from the same lack of stability as the method of centred differences in Approximation 4.1.a.

Several remedies have been proposed to obtain better FEM approximations, where the emphasis is on singularly perturbed problems like

$$a(u,v) = \int_0^1 (\varepsilon\, u'v' + b(x)u'v)\,dx = 0 , \quad u \in U,\ \forall v \in V,$$

with $\varepsilon > 0$ a "small" parameter and assuming $|b(x)| \geq b_0 = $ const > 0, $x \in \Omega$. We shall briefly review some of them without going into detail.
A first possibility is the use of artificial diffusion in the Galerkin method, i.e. to solve the problem for a value ε which is increased up to $O(h)$. This approach has a severe disadvantage because it can be only of first order accuracy and the loss of accuracy becomes particularly apparent in the boundary layers.
Second, to obtain higher order methods, the adaption of the trial- and test-spaces to the characteristics of the problems has been

proposed, which leads to the application of the Galerkin-Petrov
method. For example, in the case $|b(x)| \geq b_o > 0$, $x \in \Omega$, a boundary
layer appears only at one of the ends of Ω and the boundary layer
is of the exponential type. Therefore, in order to fit the solution
by an element of the trial-space U_h, the inclusion of exponential
trial-functions is a natural procedure, [37].

To obtain good pointwise approximations at the nodal points, it is
advantageous to adapt the test-space V_h. There are many papers devoted to this problem.

In studying nonsingular M-matrices in FEM approximations, we proceed
to consider an extension of Problem 4.8.

For this purpose, we define the bilinear functional

$$\tilde{a}(u,v) \ = \ a(u,v) \ + \ \int_0^1 q(x)\, u\, v\, dx \ , \qquad (4.106)$$

$q(x) \geq 0$, $x \in \Omega$, $a(u,v)$ given by Problem 4.8., and the linear
functional

$$l(v) \ = \ \int_0^1 f(x)\, v\, dx \ . \qquad (4.107)$$

Problem 4.8.$'$

We seek a function $u(x) \in U$ such that

$$\tilde{a}(u,v) \ = \ l(v) \ , \qquad \forall v(x) \in V$$

holds, where U,V are defined in Problem 4.8.

We assume that Problem 4.8.$'$ has a unique weak solution $u(x)$. For
instance, this is the case if $b(x) \equiv 0$, $x \in \Omega$, where $a(u,v)$ then
is symmetric, bounded and positive definite, see [4].

Let U_h, V_h be finite dimensional subspaces of U, V, respectively.
Suppose that $U_h = \mathrm{span}\left\{\varphi_i\right\}_{i=0}^{n}$, $V_h = \mathrm{span}\left\{\psi_i\right\}_{i=1}^{n-1}$ and that
$\mathrm{supp}\,\varphi_i(x) = \mathrm{supp}\,\psi_i(x) = \overline{\mathcal{E}}_i \cup \overline{\mathcal{E}}_{i+1}$, $i=1,..,n-1$, $\mathrm{supp}\,\varphi_0(x) = \overline{\mathcal{E}}_1$,
$\mathrm{supp}\,\varphi_n(x) = \overline{\mathcal{E}}_n$.

We consider the following FEM approximation to Problem 4.8.$'$.

Approximation 4.8.$'$

Find $u_h(x) \ = \ \displaystyle\sum_{i=0}^{n} y_i\, \varphi_i(x) \ \in \ U_h$ such that

$$\tilde{a}(u_h,\, \psi_j) \ = \ l(\psi_j), \qquad j=1,..,n-1,$$

holds.

For our purpose, we assume that from the partial problem

$$y_0 = u_0 ,$$
$$\sum_{i=0}^{n} y_i \, a(\varphi_i, \psi_j) = l(\psi_j) , \qquad j=1,\dots,n-1, \quad (4.108)$$
$$y_n = u_1 ,$$

there result an FEM system $Ay = f$, with A being a nonsingular tri-diagonal M-matrix, $f = (u_0, l(\psi_1), \dots, l(\psi_{n-1}), u_1)^T$.

Next we consider the influence of the terms

$$\int_0^1 q(x) \, \varphi_i \, \psi_j \, dx$$

as an additive perturbation Q on A. Additionally, let $\varphi_i(x) \geq 0$, $\psi_i(x) \geq 0$, $x \in \Omega$, for each i. Then the whole FEM system of Approximation 4.8.′ is given by

$$\tilde{A} \, y = (A + Q) \, y = f , \qquad (4.109)$$

where $Q = (q_{ji}) \geq 0$ is defined by

$$q_{ji} = \begin{cases} \int_0^1 q(x) \, \varphi_i \, \psi_j \, dx , & |i - j| \leq 1 , \\[2mm] 0 & \text{otherwise} . \end{cases} \qquad (4.110)$$

By virtue of Theorem 3.11., $\tilde{A} = A + Q$ is then a nonsingular M-matrix if all off-diagonal entries of A are nonpositive. That is, if

$$a_{ji} + q_{ji} \leq 0, \quad \text{for all } |i - j| = 1, \quad (4.111)$$

and in this case we have

$$0 \leq \tilde{A}^{-1} = (A + Q)^{-1} \leq A^{-1} .$$

If condition (4.111) fails, $\tilde{A}^{-1}$ may also exist but $\tilde{A}^{-1} \geq 0$ no longer holds, see (4.31), (4.33), (4.35) for example.

Now it is worth looking for the conditions of the region maximum principle.

Let us take a simple example. Suppose $b(x) \equiv 0$, $x \in \Omega$, in Problem 4.8.′ and choose $\psi_i = \varphi_i$ for each i with φ_i defined by (4.102). It is easily seen that

$$\int_0^1 \varphi_i \psi_j \, dx = \begin{cases} h/6 , & |i - j| = 1 , \\[2mm] 2h/3 , & i = j , \\[2mm] 0 & \text{otherwise} , \end{cases} \qquad (4.112)$$

for $j=1,\dots,n-1$.

If we use the approximate formulas

$$q_{ji} = q_j \int_0^1 \varphi_i \varphi_j \, dx \approx \int_0^1 q(x) \varphi_i \varphi_j \, dx , \quad j=1,\dots,n-1,$$

$q_j = q(x_j)$, we find

$$A = \frac{1}{h}
\begin{vmatrix}
h & 0 & & & \\
-1 & 2 & -1 & & \\
 & \ddots & \ddots & \ddots & \\
 & & -1 & 2 & -1 \\
 & & & 0 & h
\end{vmatrix}
+ \frac{h}{6}
\begin{vmatrix}
0 & 0 & & & \\
q_1 & 4q_1 & q_1 & & \\
 & \ddots & \ddots & \ddots & \\
 & & q_{n-1} & 4q_{n-1} & q_{n-1} \\
 & & & 0 & 0
\end{vmatrix} , \qquad (4.113)$$

and $\tilde{A}$ is a nonsingular M-matrix if

$$- \frac{1}{h} + \frac{q_j h}{6} \le 0 , \quad j=1,\dots,n-1 \qquad (4.114)$$

holds.

Then $y = \tilde{A}^{-1} f \ge 0$ if $f \ge 0$.

We can directly apply the region maximum principle in the case

$$q(x) \ge q_0 = \text{const} > 0, \quad x \in \Omega ,$$

because $\tilde{A}$ is then strongly row diagonally dominant, i.e., we have

$$\tilde{A} e = (1, hq_1, \dots, hq_{n-1}, 1)^T > 0 .$$

We define

$$B = \text{diag}(1, hq_1, \dots, hq_{n-1}, 1)$$

and rewrite the FEM system as

$$\tilde{A} y = B v ,$$

where

$$v = B^{-1} f = (u_0, \frac{l(\varphi_1)}{hq_1}, \dots, \frac{l(\varphi_{n-1})}{hq_{n-1}}, u_1)^T .$$

From Theorem 3.24. we deduce

$$\min\left\{ u_0, u_1, \min_{1 \le j \le n-1} \frac{l(\varphi_j)}{hq_j} \right\} \le y_i \le \max\left\{ u_0, u_1, \max_{1 \le j \le n-1} \frac{l(\varphi_j)}{hq_j} \right\}, \qquad (4.115)$$

for $i=0,\dots,n$.

A partial result now follows from the assumptions

$$f(x) \le 0, \ x \in \Omega \quad \text{and} \quad \max\{u_0, u_1\} \ge 0 .$$

Then inequality (4.115) implies

$$y_i \le \max\{u_0, u_1\} \quad \text{for } i=0,\dots,n ,$$

which is the reflection of the maximum principle of Problem 4.2.

4.4.2. Finite element approximations to two-dimensional elliptic boundary value problems

For the model problems under consideration let $\Omega \subset R^2$ be a bounded convex and polygonal region.

First, consider the weak form of Dirichlet's boundary value problem for Laplace's equation.

Problem 4.9.

Search for $u \in U = \left\{ u \in W_2^1(\Omega): \; u(x) = g(x), \; x \in \partial\Omega \right\}$ with

$$a(u,v) = \int_\Omega \sum_{s=1}^{2} \frac{\partial u}{\partial x_s} \frac{\partial v}{\partial x_s} \, dx = 0 , \quad \forall v(x) \in V ,$$

$V = \overset{\bullet}{W}_2^1(\Omega)$.

If the weak solution $u(x)$ is a harmonic function, then the boundary maximum principle (4.60) holds.

Suppose now that $\bar{\Omega}$ is triangulized. For a piecewise linear approximation, we define the set of nodal points $\{x_i\}_{i \in N}$ as the set of the vertices of all triangles. As in the FDM approximation, we use the notation $\bar{\omega}_h = \{x_i\}_{i \in N}$ and split this set into $\bar{\omega}_h = \omega_h + \gamma_h$, where $\gamma_h = \{x_i \in \partial\Omega\}_{i=1}^{n_1}$ and $\omega_h = \{x_i \in \Omega\}_{i=n_1+1}^{n}$.

We apply the Galerkin method. For this purpose introduce the basic functions $\varphi_i(x) \in C^0(\bar{\Omega})$, which are piecewise linear over the triangulation of $\bar{\Omega}$ and satisfy

$$\varphi_i(x_j) = \delta_{ij} , \qquad \delta_{ij} = \begin{cases} 0 , & i \neq j , \\ 1 , & i = j . \end{cases}$$

After this, we generate the finite dimensional spaces U_h, V_h by

$$U_h = \text{span}\left\{\varphi_i(x)\right\}_{x_i \in \bar{\omega}_h} , \qquad V_h = \text{span}\left\{\varphi_i(x)\right\}_{x_i \in \omega_h} .$$

The FEM approximation to Problem 4.9. will be determined by the following problem.

Approximation 4.9.

We seek $u_h(x) = \sum_{s=1}^{n} y_s \, \varphi_s(x) \in U_h$ with $y_s = g(x_s)$, $s=1,..,n_1$

such that

$$a(u_h, \varphi_j) = 0 , \quad j=n_1+1,\ldots,n$$

holds.

Approximation 4.9. yields an FEM system $Ay = f$, where $y=(y_1,...,y_n)^T$, $f = (g(x_1),...,g(x_n),0,...,0)^T$ and the system matrix, the so-called stiffnes matrix, is of the form (3.16).

Next we describe how to compute the nonzero entries of the sparce matrix A. To do this, it will be useful to derive an auxiliary result concerning an arbitrary triangle Δ_{ijk} of the given triangulation with the vertices x_i, x_j, $x_k \in \overline{\omega}_h$,

$$x_i = (x_{i1}, x_{i2}) , \qquad x_j = (x_{j1}, x_{j2}) , \qquad x_k = (x_{k1}, x_{k2}) ,$$

see Figure 4.11.

Fig.4.11.

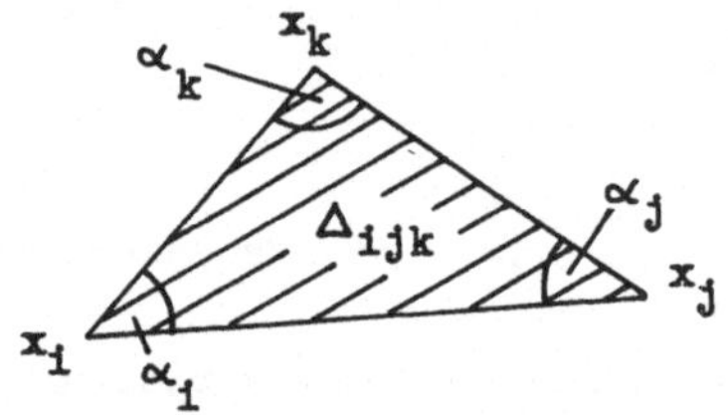

Because of $\Delta_{ijk} \subset \text{supp } \varphi_s(x)$, $s = i,j,k$, suppose that the restrictions of the basic functions $\varphi_s(x)$ on Δ_{ijk} take the form

$$\varphi_s(x)\Big|_{\Delta_{ijk}} = a_s x_1 + b_s x_2 + c_s , \quad s = i,j,k. \quad (4.116)$$

Let

$$\mathcal{A}_{ijk} = \begin{vmatrix} x_{i1} & x_{i2} & 1 \\ x_{j1} & x_{j2} & 1 \\ x_{k1} & x_{k2} & 1 \end{vmatrix} , \qquad (4.117)$$

then it is well-known from analytic geometry that

$$\left| \det \mathcal{A}_{ijk} \right| = 2 \text{ mes } \Delta_{ijk} > 0 ,$$

if the points x_i, x_j, x_k are not contained in a line and

$$\det \mathcal{A}_{ijk} > 0 ,$$

if the vertices of Δ_{ijk} are numbered anti-clockwise, as shown in Figure 4.11. This will be assumed henceforth.

We are now ready to compute the coefficients (a_s, b_s, c_s) for $s=i,j,k$ as the unique solution of the linear systems

$$\mathcal{A}_{ijk} \begin{vmatrix} a_i & a_j & a_k \\ b_i & b_j & b_k \\ c_i & c_j & c_k \end{vmatrix} = \begin{vmatrix} 1 & 0 & 0 \\ 0 & 1 & 0 \\ 0 & 0 & 1 \end{vmatrix} . \qquad (4.118)$$

Thus, we have

$$\begin{vmatrix} a_i & a_j & a_k \\ b_i & b_j & b_k \\ c_i & c_j & c_k \end{vmatrix} = \frac{1}{\det \mathcal{A}_{ijk}} \begin{vmatrix} x_{j2}-x_{k2} & x_{k2}-x_{12} & x_{12}-x_{j2} \\ x_{k1}-x_{j1} & x_{11}-x_{k1} & x_{j1}-x_{11} \\ c_i^- & c_j^- & c_k^- \end{vmatrix}, \qquad (4.119)$$

where we need only a_s and b_s in the sequel because

$$\frac{\partial \varphi_s}{\partial x_1}\bigg|_{\Delta_{ijk}} = a_s, \qquad \frac{\partial \varphi_s}{\partial x_2}\bigg|_{\Delta_{ijk}} = b_s, \qquad s=i,j,k. \quad (4.120)$$

Therefore, we shall not compute the entries c_s^- for $s=i,j,k$.

We are now able to describe that part of A which arises from $a(u_h,\varphi_s)$ for $s=i,j,k$, over the triangles Δ_{ijk}, $x_s \in \overline{\omega}_h$. It is not hard to see that this is the symmetric matrix

$$A_{ijk} = \text{mes } \Delta_{ijk} \begin{vmatrix} a_i^2 + b_i^2 & a_i a_j + b_i b_j & a_i a_k + b_i b_k \\ a_i a_j + b_i b_j & a_j^2 + b_j^2 & a_j a_k + b_j b_k \\ a_i a_k + b_i b_k & a_j a_k + b_j b_k & a_k^2 + b_k^2 \end{vmatrix} \begin{matrix} -i \\ -j \\ -k \end{matrix} . \quad (4.121)$$
$$\hspace{6cm} \begin{matrix} y_i & y_j & y_k \end{matrix}$$

From (4.119) it follows immediately that $\text{mes } \Delta_{ijk} > 0$ is equivalent to $a_s^2 + b_s^2 > 0$, $s = i,j,k$.

Further, we observe that $A_{ijk}e = 0$ holds, $e = (1,1,1)^T$, which implies $\det A_{ijk} = 0$ for each triangle Δ_{ijk}. This property comes directly from (4.119) because of

$$a_i + a_j + a_k = 0 \quad \text{and} \quad b_i + b_j + b_k = 0 .$$

The crucial point in our considerations now is to find out under which conditions the off-diagonal entries of each A_{ijk} are non-positive.

If we consider the off-diagonal entries of A_{ijk}, then it is seen that they are, except for a positive factor, the usual scalar products of the vectors which represent the sides of the triangles Δ_{ijk}. For example, we have

$$a_i a_j + b_i b_j = -\,\varkappa_{ij}((x_{12}-x_{k2})(x_{j2}-x_{k2}) + (x_{11}-x_{k1})(x_{j1}-x_{k1}))$$

$$= -\,\varkappa_{ij}\,(x_i - x_k, x_j - x_k)$$

$$= -\,\varkappa_{ij}\,\|x_i - x_k\|_2\,\|x_j - x_k\|_2\,\cos \alpha_k ,$$

with $\varkappa_{ij} > 0$. By analogy, we find

$$a_i a_k + b_i b_k = - \varkappa_{ik} \| x_i - x_j \|_2 \, \| x_k - x_j \|_2 \cos \alpha_j \, , \quad \varkappa_{ik} > 0,$$

$$a_j a_k + b_j b_k = - \varkappa_{jk} \| x_j - x_i \|_2 \, \| x_k - x_i \|_2 \cos \alpha_i \, , \quad \varkappa_{jk} > 0.$$

Thus we conclude that the off-diagonal entries of A_{ijk} are nonpositive if and only if Δ_{ijk} is not obtuse-angled. Hence, $A_{ijk} = A_{ijk}^T \in \mathscr{L}_c^{3\times3}$, see Section 3.2.

We can now assemble the FEM matrix A. Set $\hat{A}_{ijk} = (A_{ijk})_{n\times n}$, that is the entries of A_{ijk} given by (4.121) are placed in the $n\times n$ matrix $\hat{A}_{ijk}$ such that they coincide with the numbering of the nodal points $\overline{\omega}_h$, which is also expressed in the series of the components of the vector $y = (y_1,\dots,y_n)^T$. It follows that $\hat{A}_{ijk} = \hat{A}_{ijk}^T \in \mathscr{L}_r^{n\times n}$ for each triangle Δ_{ijk} and thus

$$\sum_{\text{all } \Delta_{ijk}} \hat{A}_{ijk} \in \mathscr{L}_r^{n\times n} \qquad \text{with} \qquad \left(\sum_{\text{all } \Delta_{ijk}} \hat{A}_{ijk} \right) e = 0 \, .$$

The whole matrix A is then given by

$$A = \begin{pmatrix} I & 0 \\ 0 & 0 \end{pmatrix} + \sum_{\text{all } \Delta_{ijk}} \hat{A}_{ijk} = \begin{pmatrix} I & 0 \\ A_{21} & A_{22} \end{pmatrix}, \qquad (4.122)$$

I the unit matrix of order n_1.

We are now going to discuss some properties of A.

The matrix A is an L-matrix, since all the diagonal entries are positive and Δ_{ijk} are not obtuse-angled, i.e., the off-diagonal entries are nonpositive. By $(A_{21} \, A_{22}) e = 0$ and $A_{21} \le 0, \neq 0$, we find that A_{22} is a weakly row diagonally dominant L-matrix, which is also irreducible because the associated directed graph of A_{22} is strongly connected, see Section 4.2.

Thus A_{22} is a nonsingular M-matrix by Theorem 3.12. and such is A by virtue of Theorem 3.19. Finally, we also find that $A_{21}^T e'' < 0'$ holds. Hence, by Theorem 3.21., the boundary maximum principle holds. Therefore, we have

$$\min_{x_j \in \gamma_h} \left\{ g(x_j) \right\} \le y_i \le \max_{x_j \in \gamma_h} \left\{ g(x_j) \right\} \, , \qquad \forall i \in N. \qquad (4.123)$$

We remark that if some Δ_{ijk} of the triangulation is obtuse-angled, then, the M-matrix property of A must not fail in any case. To illustrate this, let the angle α_k in Figure 4.11. be obtuse.

Then $a_i a_j + b_i b_j > 0$ in (4.121). Suppose that Δ_{ilj} is the adjacent triangle, see Figure 4.12.

Fig.4.12.

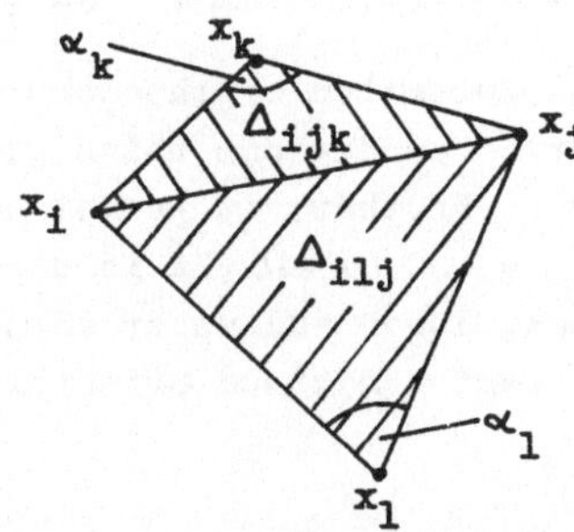

Then the entry a_{ij} of A is the sum of the expressions of the form $a_i a_j + b_i b_j$ which come from Δ_{ijk} and Δ_{ilj} separately, and it may happen that $a_{ij} \leq 0$ if one of the angles α_k or α_l is obtuse. Here, we shall not further analyse such situations, see [12].

Finally, we briefly look at an extension of Problem 4.9.

Problem 4.9.´

Search for $u \in U$ such that

$$\tilde{a}(u,v) \;=\; a(u,v) \;+\; \int_\Omega q(x)\, u\, v\, dx \;=\; 1(v)\,, \quad \forall v \in V$$

holds.

Suppose that $a(u,v)$, U, V are given by Problem 4.9. Further, let $q(x) \geq 0$, $x \in \Omega$ and $1(v) = \int_\Omega f(x)\, v\, dx$. We assume that Problem 4.9.´ has a unique weak solution $u(x) \in U$.

Applying the Galerkin method to Problem 4.9.´, we get the following FEM approximation.

Approximation 4.9.´

Let U_h, V_h be the same finite dimensional spaces as in Approximation 4.9. Then, we seek for $u_h(x) = \sum_{i=1}^{\tilde{n}} y_i \varphi_i(x) \in U_h$ with $y_s = g(x_s)$, $s = 1, \dots, n_1$, such that

$$\tilde{a}(u_h, \varphi_j) \;=\; 1(\varphi_j)\,, \quad j = n_1 + 1, \dots, n$$

holds.

The resulting FEM system is then of the form $\tilde{A}y = (A + Q)y = \tilde{f}$, where A comes from Approximation 4.9., $Q = (q_{ij}) \geq 0$ is determined by the terms

$$q_{ji} \;=\; \int_\Omega q(x)\, \varphi_i\, \varphi_j\, dx$$

and $\quad \widetilde{f} = (g(x_1),\ldots,g(x_{n_1}),f_{n_1+1},\ldots,f_n)^T \quad$ with

$$f_j = \int_\Omega f(x)\,\varphi_j\,dx\ , \quad j=n_1+1,\ldots,n.$$

In view of the additive perturbation of the nonsingular M-matrix A, i.e. $\widetilde{A} = A + Q$, we can derive conditions which guarantee that A is also a nonsingular M-matrix. Further, it is not hard to see that under the assumption $q(x) \geqslant q_0 > 0$, $x \in \Omega$, $\widetilde{A}$ is strongly diagonally dominant of its rows. Hence, lower and upper solution bounds are directly available by the region maximum principle, see Section 3.4.2.

4.5. Method of lines

In the present section we return to the linear time-dependent problems of the parabolic type, which are to be solved approximately by method of lines, i.e. by semi-discretization methods.
In general, we distinguish two principal approaches, viz., the longitudinal and the transversal methods of lines. The first one assumes discretization in space from which a system of ordinary differential equations with a well-defined initial condition results, i.e. a Cauchy problem, see [17] and elsewhere. The second one is known as the Method of Rothe, which produces, after a priori time discretization, a series of elliptic problems to be solved. For analytic aspects of this approach, the reader is referred to [15].

Our main interest here concerns the longitudinal method. For brevity, we consider one-dimensional linear parabolic problems only and focus our attention on an intimate connection between nonsingular M-matrices and the solution of the arising Cauchy problems.

We shall begin by reviewing some results related to ordinary linear differential equation systems with constant data.
Consider the Cauchy problem

$$\begin{aligned}
\dot{y} + A y &= f\ , \quad t > 0, \\
y(0) &= y_0,
\end{aligned} \tag{4.124}$$

where

$$y = y(t) = (y_1(t),\ldots,y_n(t))^T\ ,$$

$A = (a_{ij})$ is a constant $n{\times}n$ matrix ,

$y_0 \in R^n$ the initial value ,

$f \in R^n$ a time-independent vector .

110

In our considerations we take A and f to be a constant matrix and
a constant vector to get the solution y(t) of the Cauchy problem
(4.124) explicitly. That is, if A is a nonsingular matrix, then the
unique solution is given by

$$y(t) = A^{-1}f + \exp(-tA)(y_0 - A^{-1}f) \quad \text{for all } t \geqslant 0. \qquad (4.125)$$

Additionally, for M-matrices A we have the following statement.
Let A be a nonsingular M-matrix. Furthermore, assume that $f \geqslant 0$,
$Ay_0 \geqslant f$. Then the Cauchy problem (4.124) has a unique nonnegative
solution, i.e. $y(t) \geqslant 0$ for all $t \geqslant 0$, with

$$\lim_{t \to \infty} y(t) = A^{-1}f \geqslant 0 , \qquad (4.126)$$

and the solution y(t) is uniformly bounded for all $t \geqslant 0$.

For the proof we cite [26]. Let A be a nonsingular M-matrix, then
the unique solution of (4.124) is given by formula (4.125). By our
assumptions, all the solution terms on the right-hand side are non-
negative. That is, we have $A^{-1}f \geqslant 0$, further, $Ay_0 \geqslant f \geqslant 0$ implies
$y_0 - A^{-1}f \geqslant 0$, and since -tA is essentially positive for all $t > 0$
by Property 3.9., we get $\exp(-tA) > 0$, $t > 0$, see Proposition 1.46.
Thus $y(t) \geqslant 0$ for all $t > 0$ with $y(0) = y_0 \geqslant 0$.
To complete the proof, we show that y(t) is uniformly bounded for
all $t \geqslant 0$. Assume that the nonsingular M-matrix A is represented as

$$A = sI - B \quad \text{with} \quad s > 0, \ B \geqslant 0, \ s > S(B) .$$

Thus

$$\exp(-tA) = \exp(-stI + tB) = \exp(-st)\exp(tB) ,$$

and

$$\| \exp(-tA) \|_2 = \exp(-st) \| \exp(tB) \|_2 = \exp(-st)\exp(tS(B))$$

$$= \exp(-t(s-S(B))) \xrightarrow[t \to \infty]{} 0 .$$

Therefore

$$\lim_{t \to \infty} \exp(-tA) = 0 ,$$

implies

$$\lim_{t \to \infty} y(t) = A^{-1}f , \qquad (4.127)$$

the steady-state solution of the problem.

Furthermore, we find

$$\|y(t)\|_2 \;\leq\; \|A^{-1}f\|_2 \;+\; \|\exp(-tA)\|_2\,\|y_0 - A^{-1}f\|_2$$

$$= \|A^{-1}f\|_2 \;+\; \exp(-t(s-S(B)))\,\|y_0 - A^{-1}f\|_2 \qquad (4.128)$$

$$\leq \|A^{-1}f\|_2 \;+\; \|y_0 - A^{-1}f\|_2\,,$$

and the assertion is proved.

The last inequality implies for $f = 0$

$$\|y(t)\|_2 \;\leq\; \exp(-t(s-S(B)))\,\|y_0\|_2, \qquad (4.129)$$

that is, the Euclidian norm of the solution $y(t)$ is monotone decreasing for $t \longrightarrow \infty$ and (4.129) also yields

$$\lim_{t \longrightarrow \infty} y(t) \;=\; 0\,.$$

If the nonsingular M-matrices A arise from discretization methods, then we get stiff Cauchy problems, which coincide with condition numbers $\operatorname{cond} A \gg 1$, see [17] and elsewhere.

Now just as we have done before, we consider a model problem which is to be solved approximately by longitudinal methods of lines.

<u>Problem 4.10.</u>   Let $\Omega = (0,1)$. Consider

$$\frac{\partial u}{\partial t} \;+\; \tilde{L}\,u \;=\; \tilde{f}(x)\,, \qquad (x,t)\in\Omega\times(0,\infty)\,,$$

$$u(x,0) \;=\; u_0(x)\,, \qquad x\in\Omega\,,$$

$$\left(\alpha_0 u - \beta_0\frac{\partial u}{\partial x}\right)\Big|_{x=0} \;=\; g_0\,,$$

$$\left(\alpha_1 u + \beta_1\frac{\partial u}{\partial x}\right)\Big|_{x=1} \;=\; g_1\,, \qquad 0 < t < \infty\,.$$

For simplicity, suppose that L is an elliptic differential operator with constant coefficients k, b, q, i.e.

$$\tilde{L}\,u \;=\; -k\frac{\partial^2 u}{\partial x^2} \;+\; b\frac{\partial u}{\partial x} \;+\; q\,u\,,$$

where $k > 0$, $q \geqslant 0$. Additionally, let α_s, β_s, g_s also be constant with $\alpha_s > 0$, $\beta_s \geqslant 0$, $s = 0,1$.

Furthermore, it is assumed here that Problem 4.10. has a unique classical solution $u(x,t)$.

It is well known that the parabolic Problem 4.10. obeys a maximum principle, involves a conservation law and its solution tends to a steady-state solution $u(x)$ for $t \longrightarrow \infty$, i.e. to a equilibrium or time-independent solution. Since such results are classical, we shall not formulate these properties in detail and refer the reader to [22,24].

First, we consider the method of lines for FDM approximation in space.

For this and for later considerations, assume a uniform grid $\overline{\omega}_h$ on $\overline{\Omega}$, $\overline{\omega}_h = \{ x_i : x_i = ih, \ i=0,\ldots,n \}$, with step size $h = 1/n$.

Now denote

$$y_i(t) \ = \ y(x_i,t) \ \approx \ u(x_i,t) \ , \quad i=0,\ldots,n \ , \quad t \geqslant 0. \qquad (4.130)$$

Using three-point difference schemes to approximate $\widetilde{L}u$ on ω_h in analogy to Approximations 4.1.a-c, we derive difference equations of the form

$$\dot{y}_i(t) \ + \ \widetilde{L}_h \, y_i(t) \ = \ \widetilde{f}_i \ , \quad i=1,\ldots,n-1 \ , \qquad (4.131)$$

where $\widetilde{L}_h$ contains constant coefficients for fixed h, setting $\widetilde{f}_i = \widetilde{f}(x_i)$.

In order to approximate the boundary conditions, we set

$$(\alpha_0 \ + \ \beta_0/h) \, y_0(t) \ - \ \beta_0/h \, y_1(t) \ = \ g_0 \ ,$$
$$-\alpha_1/h \, y_{n-1}(t) \ + \ (\alpha_1 + \beta_1/h) \, y_n(t) \ = \ g_1 \ , \qquad (4.132)$$

such that we can eliminate $y_0(t)$ and $y_n(t)$ in (4.131) via

$$y_0(t) \ = \ (g_0 + \beta_0/h \, y_1(t))/(\alpha_0 + \beta_0/h) \ ,$$
$$y_n(t) \ = \ (g_1 + \beta_1/h \, y_{n-1}(t))/(\alpha_1 + \beta_1/h) \ . \qquad (4.133)$$

Hence, we can rewrite (4.131) as an ordinary differential equation system, the approximation to Problem 4.10.

<u>Approximation 4.10.a</u>

Let $y = y(t) = (y_1(t),\ldots,y_{n-1}(t))^T$. Then

$$\dot{y} \ + \ A \, y \ = \ f \ , \quad t > 0 \ ,$$
$$y_0 = (u_0(x_1),\ldots,u_0(x_{n-1}))^T \ , \qquad (4.134)$$

with a constant tridiagonal matrix A and a constant vector f given

by $f = (\tilde{f}_1,\ldots,\tilde{f}_{n-1})^T$. The initial condition y_0 comes from the
initial condition of Problem 4.10.

Our essential assumption is now that A will be a nonsingular
M-matrix.

Thus, Approximation 4.10.a states the Cauchy problem in order to
approximate the parabolic Problem 4.10. Its unique solution y(t) is
of the form (4.125), exhibits asymptotic behaviour (4.126) and sa-
tisfies the estimate (4.128). Furthermore, $f \geqslant 0$ and $y_0 \geqslant A^{-1}f$ guarantee $y(t) \geqslant 0$ for $t \geqslant 0$. But as will be seen in the following example,
the latter condition is of course not necessary for $y(t) \geqslant 0$, $t \geqslant 0$.

We look briefly at an example.

Example 4.10.

$$\frac{\partial u}{\partial t} \;-\; \frac{\partial^2 u}{\partial x^2} \;=\; \tilde{f}(x) \;, \qquad (x,t) \in \Omega \times (0,\infty) \;,$$

$$u(x,0) \;=\; u_0(x) \;, \qquad x \in \Omega \;,$$

$$u(0,t) = g_0 \;, \quad u(1,t) = g_1 \;, \qquad 0 < t < \infty \;.$$

The steady-state solution u(x) is then defined by

$$- u'' = \tilde{f}(x) \;,$$
$$u(0) = g_0 \;, \qquad u(1) = g_1 \;. \tag{4.135}$$

Applying Approximation 4.10.a to Example 4.10., we find that the
nonsingular M-matrix A is taken by

$$A \;=\; \frac{1}{h^2} \, \mathrm{tridiag}(-1,2,-1) \;,$$

and the right-hand side vector is of the form

$$f \;=\; \frac{1}{h^2} \, (g_0,0,\ldots,0,g_1)^T \;+\; (\tilde{f}_1,\ldots,\tilde{f}_{n-1})^T.$$

According to (4.125), the conditions $f \geqslant 0$, $y_0 \geqslant A^{-1}f$ are sufficient
for $y(t) \geqslant 0$, $t \geqslant 0$.

To obtain a deeper insight into the meaning of the condition $y_0 \geqslant A^{-1}f$, let $\tilde{f}(x) \equiv 0$, $x \in \Omega$. In this case the maximum principle for
Example 4.10. yields

$$\min\left\{g_0,g_1, \min_{x \in \Omega} u_0(x)\right\} \leqslant u(x,t) \leqslant \max\left\{g_0,g_1, \max_{x \in \Omega} u_0(x)\right\} \;, \tag{4.136}$$

for each $(x,t) \in \Omega \times (0,\infty)$ and $u(x) = g_1 x + g_0(1-x)$, $x \in \Omega$, is the
steady-state solution. Hence, $f \geqslant 0$ is equivalent to $\min\{g_0,g_1\} \geqslant 0$,
but $y_0 \geqslant A^{-1}f$ implies $u_0(x) \geqslant u(x)$, $x \in \Omega$, which is not necessary.

This property follows from Example 3.34., where $A^{-1} = (a_{ij}^-)$ is explicitly given and thus

$$A^{-1}f = \frac{g_0}{n}(n-1,n-2,\ldots,1)^T + \frac{g_1}{n}(1,\ldots,n-2,n-1)^T =$$

$$= (u(x_1),\ldots,u(x_{n-1}))^T.$$

Adding up all the equations of $\dot{y} + Ay = f$, we get a discrete version of the conservation law of the parabolic problem.

A second approach of the method of lines uses FEM approximations in space. It requires a weak formulation of the parabolic problems under consideration, see [13].
According to Problem 4.10., we omit here the weak form and turn directly to the semi-discrete FEM approximation.
For simplicity, let us again assume the boundary conditions

$$u(0,t) = g_0 \quad \text{and} \quad u(1,t) = g_1 , \quad t > 0 , \qquad (4.137)$$

where g_0 and g_1 are constants.
Suppose that over a uniform subdivision of Ω into finite elements $\mathcal{E}_i = (x_{i-1},x_i)$, $i \in N$ with $x_i - x_{i-1} = h = 1/n$ the piecewise linear basic functions $\varphi_i(x) \in C(\bar{\Omega})$, defined by (4.102), are used to the generation of approximate FEM solutions. Furthermore, let U_h, V_h be the same finite dimensional spaces as in Approximation 4.8.
Denote by

$$(u,v) = \int_0^1 u\,v\,dx , \quad \text{scalar product} ,$$

$$a(u,v) \quad \text{bilinear functional generated by the elliptic part}$$
$$\text{of the parabolic differential equation} ,$$

$$l(v) = \int_0^1 f(x)\,v\,dx , \quad \text{linear functional} ,$$

for $\forall u \in U_h$, $\forall v \in V_h$.

Then, the approximation to Problem 4.10. with boundary conditions (4.137) is stated as follows, see [13,25].

Approximation 4.10.b

Seek $u_h(x,t) = \sum_{i=0}^{n} y_i(t)\,\varphi_i(x) \in U_h$ with $y_0(t) = g_0$, $y_n(t) = g_1$

and $y_i(0) = u_0(x_i)$, $i=1,\ldots,n-1$ such that

$$\left(\frac{\partial u_h}{\partial t},\varphi_j\right) + a(u_h,\varphi_j) = l(\varphi_j), \quad j=1,\ldots,n-1$$

holds for all $t > 0$.

From this weak form there results a system of ordinary differential equations. That is,

$$\sum_{i=1}^{n-1} \dot{y}_i(t)\,(\varphi_i,\varphi_j) \;+\; \sum_{i=0}^{n} y_i(t)\,a(\varphi_i,\varphi_j) \;=\; l(\varphi_j), \qquad (4.138)$$

for $j=1,..,n-1$, where $y_0(t) = g_0$ and $y_n(t) = g_1$ are to be incorporated. Then Approximation 4.10.b implies the following Cauchy problem via (4.138)

$$\begin{aligned} M\,\dot{y} \;+\; A\,y \;&=\; f\,, \qquad t > 0\,, \\ y(0) \;&=\; y_0\,. \end{aligned} \qquad (4.139)$$

The initial value y_0 is defined by (4.134).

In our considerations we again assume that $A = (a_{ji})$, $a_{ji} = a(\varphi_i,\varphi_j)$ is a nonsingular M-matrix, see also Section 4.4. For the special Problem 4.10., the matrix A is constant and tridiagonal, which ceases for more general problems.

The right-hand side vector f takes the form

$$f \;=\; (l(\varphi_1),\dots,l(\varphi_{n-1}))^T \;+\; g\,,$$

where g contains g_0 and g_1.

Let us now focus attention on $M = (m_{ji})$, the so-called mass matrix, see [13]. Its entries are given by (4.112). Thus, we have

$$M \;=\; \frac{h}{6}\,\text{tridiag}(1,4,1) \;\geq\; 0\,. \qquad (4.140)$$

The mass matrix M is strongly diagonally dominant, which implies $\det M \neq 0$. Additionally, we find that $\det M > 0$, since the diagonal entries are all positive. The determinant of M can be computed iteratively by formula (3.12). Furthermore, from (4.35) we find that the sign of the entries of M^{-1} have a chess board pattern. That is

$$M^{-1} \;=\; ((-1)^{j+i}\, m^{-}_{ji}) \quad \text{with} \quad m^{-}_{ji} > 0\,, \qquad (4.141)$$

for all $i,j = 1,..,n-1$.

For example

$$\begin{pmatrix} 4 & 1 & 0 \\ 1 & 4 & 1 \\ 0 & 1 & 4 \end{pmatrix}^{-1} = \frac{1}{56}\begin{pmatrix} 15 & -4 & 1 \\ -4 & 16 & -4 \\ 1 & -4 & 15 \end{pmatrix},\quad \begin{pmatrix} 4 & 1 & & \\ 1 & 4 & 1 & \\ & 1 & 4 & 1 \\ & & 1 & 4 \end{pmatrix}^{-1} = \frac{1}{209}\begin{pmatrix} 56 & -15 & 4 & -1 \\ -15 & 60 & -16 & 4 \\ 4 & -16 & 60 & -15 \\ -1 & 4 & -15 & 56 \end{pmatrix}$$

$$
\begin{vmatrix} 4 & 1 & & & \\ 1 & 4 & 1 & & \\ & 1 & 4 & 1 & \\ & & 1 & 4 & 1 \\ & & & 1 & 4 \end{vmatrix}^{-1} = \frac{1}{780} \begin{vmatrix} 209 & -56 & 15 & -4 & 1 \\ -56 & 224 & -60 & 16 & -4 \\ 15 & -60 & 225 & -60 & 15 \\ -4 & 16 & -60 & 224 & -56 \\ 1 & -4 & 15 & -56 & 209 \end{vmatrix} ,
$$

from which it is clear how to proceed.

Now we consider the Cauchy problem (4.139) in the explicit form

$$
\dot{y} + M^{-1}A\,y = M^{-1}f , \quad t > 0 ,
$$
$$
y(0) = y_0 . \tag{4.142}
$$

We can readily see that $M^{-1}A$ is a monotone matrix, since it is
invertible and $(M^{-1}A)^{-1} = A^{-1}M \geqslant 0$.

Then the unique solution $y(t)$, $t \geqslant 0$ of the Cauchy problem (4.142) is
also given by formula (4.125), putting $A := M^{-1}A$. But $M^{-1}A$ is generally no longer an M-matrix as will be seen from the following simple
example.

For this purpose, let us again consider Example 4.10. In this case
we have

$$
a(u,v) = \int_0^1 u'\,v'\,dx ,
$$

and the Cauchy problem (4.139) takes the form

$$
\frac{h}{6} \begin{vmatrix} 4 & 1 & & & \\ 1 & 4 & 1 & & \\ & & \ddots & & \\ & & 1 & 4 & 1 \\ & & & 1 & 4 \end{vmatrix} \dot{y} + \frac{1}{h} \begin{vmatrix} 2 & -1 & & & \\ -1 & 2 & -1 & & \\ & & \ddots & & \\ & & -1 & 2 & -1 \\ & & & -1 & 2 \end{vmatrix} y = \begin{vmatrix} 1(\varphi_1) \\ \vdots \\ \vdots \\ 1(\varphi_{n-1}) \end{vmatrix} + \frac{1}{h} \begin{vmatrix} g_0 \\ 0 \\ \vdots \\ 0 \\ g_1 \end{vmatrix} \tag{4.143}
$$

for $t > 0$ with the initial condition $y(0) = y_0$.

It is now not hard to see that $M^{-1}A$ is a matrix with a chess board
sign pattern of its entries which are all nonzero. Hence, symboli-
cally written as

$$
M^{-1}A = \begin{vmatrix} + & - & + & - & \cdots & \\ - & + & - & + & & \\ + & - & + & - & & \\ - & + & - & + & & \\ \vdots & & & & \ddots & \\ & & & & & + \end{vmatrix} .
$$

A detailed analysis of a situation like this is necessary but beyond
the scope of our book.

To overcome the difficulties which arise in Approximation 4.10.b ,
the so-called lumped mass type approximation has been developed,
see [13] and elsewhere. This approach to parabolic problems may be
interpreted as a multi-base FEM approximation, where different basic

function systems are used to approximate the different terms of the
weak formulation of the problem under consideration.

In the following, we shall briefly illustrate the main ideas of this
method applied to Problem 4.10.

For this purpose, let $\overline{\omega}_h = \{x_i\}_{i=0}^{n}$ be a uniform set of nodal points
in $\overline{\Omega}$. Next we define elementary regions

$$\mathcal{H}_i = (x_i - h/2 , x_i + h/2) \quad \text{for } i=1,..,n-1,$$

and introduce the system of basic functions $\psi_i(x)$, $i=1,..,n-1$,
where each $\psi_i(x)$ is the characteristic function of the elementary
region $\mathcal{H}_i$, that is

$$\psi_i(x) = \begin{cases} 1 & \text{for } x \in \mathcal{H}_i , \\ 0 & \text{otherwise .} \end{cases} \tag{4.144}$$

Then the following lumped mass FEM approximation to Problem 4.10.
will be considered.

Approximation 4.10.c

Let all the notations be the same as in Approximation 4.10.b .

Then seek $u_h(x,t) = \sum_{i=0}^{n} y_i(t)\,\varphi_i(x) \in U_h$ with $y_0(t) = g_0$,

$y_n(t) = g_1$, and $y_i(0) = u_0(x_i)$, $i=1,..,n-1$ such that

$$\sum_{i=1}^{n-1} \dot{y}_i(t)\,(\psi_i, \psi_j) + a(u_h, \varphi_j) = l(\varphi_j)$$

holds for $j=1,..,n-1$ and for each $t > 0$.

It is then quite easy to see that $\mathcal{V} = ((\psi_i, \psi_j)) = hI$ in our model
problem. In general, $\mathcal{V}$ will be a positive diagonal matrix.

From Approximation 4.10.c we find

$$\dot{y} + \frac{1}{h} A y = \frac{1}{h} f . \tag{4.145}$$

Thus, if A is a nonsingular M-matrix, then A/h is also a nonsingular
M-matrix.

In general, we have

$$\dot{y} + \mathcal{V}^{-1} A y = \mathcal{V}^{-1} f , \quad t > 0, \tag{4.146}$$

and if A is a nonsingular M-matrix, then $\mathcal{V}^{-1}A$ also has this property, see Example 3.1.3.

It is readily seen that the system of ordinary differential equations (4.145) is equivalent to Approximation 4.10.a . In summary,
the lumped mass type FEM approximation is closely related to the
method of lines with FDM approximation in space.

5. M-MATRICES AND EIGENVALUE PROBLEMS

In this chapter we study an interesting relationship between special
linear elliptic eigenvalue problems and its FDM approximation, which
involves M-matrices. In this context, we do not take into account
FEM approximations, since they show the same qualitative properties
if they also lead to eigenvalue problems for M-matrices. The reader
interested in results concerning the convergence of eigenvalues and
their eigenvectors of the discretized problems to those of the elliptic problems is referred to [6,11,25] and elsewhere.
The main example we consider is the Sturm-Liouville eigenvalue problem in bounded regions, which arises from a number of important applications and for which an advanced theory is available, see [27].
In the simplest case, which is the eigenvalue problem for the negative Laplacian $-\Delta$, the properties we shall consider become completely obvious.

5.1. A cursory view of the Sturm-Liouville eigenvalue problem

To begin with, we state the general elliptic eigenvalue problem and
review its main properties. In doing this, we follow [27].
Let $\Omega \subset R^d$, $d \geq 1$, be a bounded region with boundary $\partial\Omega$ piecewise
smooth. We consider the following eigenvalue problem.

Problem 5.1. (Sturm-Liouville eigenvalue problem)

$$Lu = -\nabla(k\nabla u) + qu = \lambda u, \quad x \in \Omega,$$

$$\alpha u + \beta \frac{\partial u}{\partial \nu} = 0, \quad x \in \partial\Omega.$$

We seek numbers λ such that Problem 5.1. has nontrivial solutions
$u(x) \in C^2(\bar{\Omega}) \cap C^1(\bar{\Omega})$, i.e. $u(x) \not\equiv 0$, $x \in \Omega$, the corresponding
eigenfunctions.

Suppose that $k(x) \in C^1(\bar{\Omega})$, $q(x) \in C(\bar{\Omega})$ with $k(x) > 0$, $q(x) \geq 0$,
$x \in \Omega$. Further, let ν be the direction of the outward normal to Ω
and assume that $\alpha \in C(\partial\Omega)$, $\beta \in C(\partial\Omega)$ with $\alpha(x) \geq 0$, $\beta(x) \geq 0$,
$\alpha(x) + \beta(x) > 0$ for each $x \in \partial\Omega$.

Under these assumptions Problem 5.1. has the following properties.

Property 5.2.

The self-adjoint positive semidefinite or positive definite elliptic

operator L, with the homogeneous boundary conditions $\alpha u + \beta \frac{\partial u}{\partial \nu} = 0$
on $\partial \Omega$, has a real discrete spectrum $\sigma(L) \subset R^1_+$. The spectrum $\sigma(L)$
contains countably many eigenvalues and infinity is the only possible
limit point of eigenvalues $\lambda \in \sigma(L)$.

Property 5.3.

We have $\lambda = 0 \in \sigma(L)$ if and only if $q \equiv 0$, $x \in \Omega$, $\alpha \equiv 0$, $x \in \partial \Omega$ holds.
If $\lambda = 0 \in \sigma(L)$ then $\lambda = 0$ is a simple eigenvalue and $u(x) = $ const
$\neq 0$, $x \in \Omega$, is the corresponding eigenfunction.

Property 5.4.

Let $\lambda^* = \inf_{\lambda \in \sigma(L)} \lambda$. Then there exists a corresponding eigenfunction
$u^*(x)$ which is of one sign in Ω . That is, $u^*(x)$ can be chosen such
that $u^*(x) > 0$, $x \in \Omega$ holds.
We remark that for $\lambda^* = 0$ this is now obvious by Property 5.3.

Property 5.5.

For $\lambda_1, \lambda_2 \in \sigma(L)$ with $\lambda_1 \neq \lambda_2$ the corresponding eigenfunctions
u_1 and u_2 are L_2-orthogonal, that means, we have

$$(u_1,u_2) \;=\; \int_\Omega u_1 \, u_2 \, dx \;=\; 0 \; . \qquad (5.1)$$

As a direct consequence of the Properties 5.4. and 5.5. we find that
for Problem 5.1. there cannot exist two linear independent eigenfunctions, of which are of one sign in Ω .

5.2. One-dimensional Sturm-Liouville eigenvalue problems and their
     finite difference approximation

We shall begin by one of the simplest examples, for which the whole
spectrum and all eigenvectors are known. This is the following very
special Sturm-Liouville eigenvalue problem.

Problem 5.6.

Let $\Omega = (0,1)$, $0 < 1 < \infty$, $\partial \Omega = \{0,1\}$. Consider

$$- u'' \;=\; \lambda u , \qquad x \in \Omega ,$$

with one of the following boundary conditions on $\partial \Omega$:

a) homogeneous Dirichlet conditions on $\partial \Omega$

$$u(0) \;=\; u(1) \;=\; 0 ,$$

b) homogeneous Neumann and Dirichlet conditions on $\partial \Omega$

$$u'(0) \;=\; u(1) \;=\; 0 ,$$

c) homogeneous **Neumann** conditions on $\partial\Omega$

$$u'(0) \; = \; u'(1) \; = \; 0 \, .$$

The complete solution of Problem 5.6. is cited from $[28]$. All of the eigenvalues λ_k, k=1,2,.. of Problem 5.6. are simple and the eigenfunctions are normed such that $\| u_k \| = 1$ holds. The behaviour of the eigenvectors $u_k(x)$ for k=1,2,3,4 is depicted in the inserted figures.

Solution 5.6.a

$$\lambda_k \; = \; \left(\frac{k\pi}{l}\right)^2 \, , \qquad u_k(x) \; = \; \sqrt{\frac{2}{l}} \, \sin\frac{k\pi x}{l} \, , \qquad x\in\Omega \, , \qquad k=1,2,\dots$$

Fig.5.1.

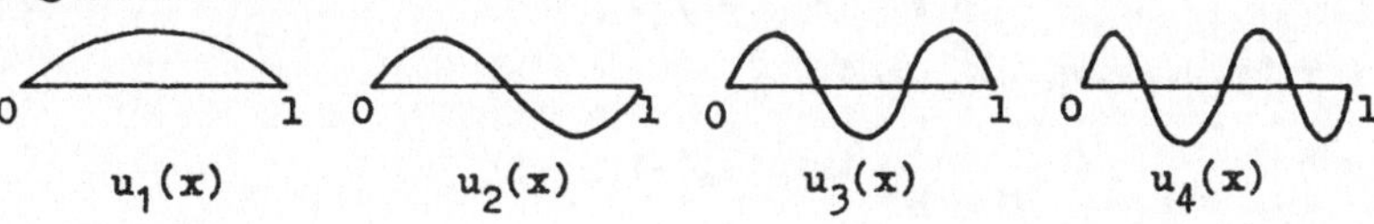

Solution 5.6.b

$$\lambda_k \; = \; \left(\frac{(k-1/2)\pi}{l}\right)^2 \, , \qquad u_k(x) \; = \; \sqrt{\frac{2}{l}} \, \cos\frac{(k-1/2)\pi x}{l} \, , \qquad x\in\Omega \, ,$$

$$k=1,2,\dots$$

Fig.5.2.

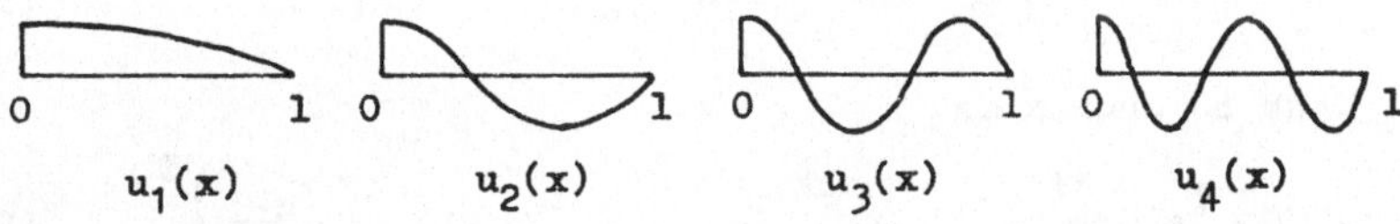

Solution 5.6.c

$$\lambda_k \; = \; \left(\frac{(k-1)\pi}{l}\right)^2 \, , \qquad u_1(x) \; = \; \sqrt{\frac{1}{l}} \, , \qquad u_k(x) \; = \; \sqrt{\frac{2}{l}} \, \cos\frac{(k-1)\pi x}{l} \, ,$$

$$x\in\Omega \, , \qquad k=2,3,\dots$$

Fig.5.3.

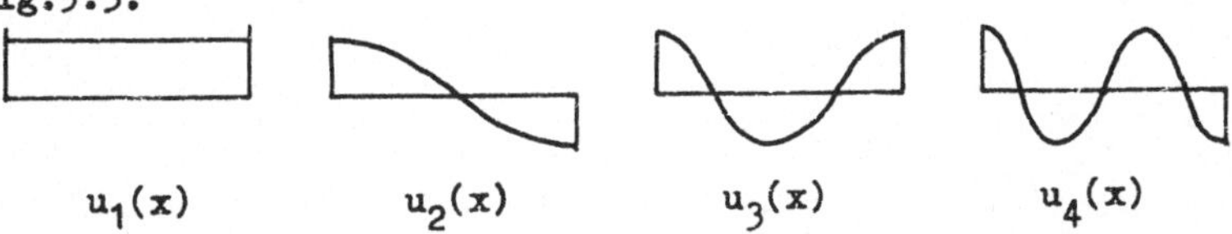

Now let us consider an FDM discretization of Problem 5.6. over a uniform grid $\overline{\omega}_h$, which is defined as follows

$$\overline{\omega}_h \; = \; \left\{ x_i = ih \, , \quad i=0,\dots,n \, , \text{ where } h=1/n \right\}. \qquad (5.2)$$

Let y_i be the approximate values of the eigenfunctions of Problem
5.6. at x_i, then we approximate the differential equation by the
series of difference equations

$$- D_+ D_- y_i \;=\; \mu\, y_i \qquad \text{for } i=1,\ldots,n-1. \qquad (5.3)$$

To complete the discrete approximations, we have to incorporate the
discretized homogeneous boundary conditions under consideration.
Thus, we derive three matrix eigenvalue problems. Each of the matrices is an M-matrix for which the whole solution is available.

Approximation 5.6.a

From (5.3), it follows with $y_0 = y_n = 0$ that

$$A\,y \;=\; \mu\,y\,,$$

where $y = (y_1,\ldots,y_{n-1})^T$ and

$$A \;=\; \frac{1}{h^2}\,\text{tridiag}(-1,2,-1)$$

is of order n-1.

By Theorem 2.22., the matrix A is a symmetric nonsingular M-matrix,
i.e. a Stieltjes matrix. From Section 2.2.3., we derive all $\mu_k \in \mathfrak{S}(A)$
and all the corresponding eigenvectors v_k , $k=1,\ldots,n-1$.
Thus

$$\mu_k \;=\; \frac{4}{h^2}\,\sin^2 \frac{k\pi}{2n}\,, \qquad\qquad (5.4)$$

and v_k can be chosen as

$$v_k \;=\; \sqrt{\frac{2}{1}}\left(\sin\frac{k\pi x_1}{1}\,,\; \sin\frac{k\pi x_2}{1}\,,\;\ldots\,,\; \sin\frac{k\pi x_{n-1}}{1}\right)^T\,, \qquad (5.5)$$

for $k=1,\ldots,n-1$.

Approximation 5.6.b

From (5.3), we have with $y_0 = y_1$ and $y_n = 0$ that

$$A\,y \;=\; \mu\,y\,,$$

where $y = (y_0,y_1,\ldots,y_{n-1})^T$ and

$$A \;=\; \frac{1}{h^2}\begin{pmatrix} 2 & -2 & & & \\ -1 & 2 & -1 & & \\ & \ddots & \ddots & \ddots & \\ & & -1 & 2 & -1 \\ & & & -1 & 2 \end{pmatrix}_{n \times n}$$

It follows from Theorem 2.22. that A is a nonsingular M-matrix and
it needs a little computation to see that $\mu_k \in \sigma(A)$ are given by

$$\mu_k = \frac{4}{h^2} \sin^2 \frac{(k-1/2)\pi}{2n} \, , \tag{5.6}$$

and v_k can be chosen as

$$v_k = \sqrt{\frac{2}{1}} \left(1, \cos\frac{(k-1/2)\pi x_1}{1} , \ldots , \cos \frac{(k-1/2)\pi x_{n-1}}{1} \right)^T , \tag{5.7}$$

for $k=1,\ldots,n$.

<u>Approximation 5.6.c</u>

It follows from (5.3) with $y_0 = y_1$ and $y_{n-1} = y_n$ that

$$A y = \mu y ,$$

where $y = (y_0, y_1, \ldots, y_{n-1}, y_n)^T$ and

$$A = \frac{1}{h^2} \begin{pmatrix} 2 & -2 & & & \\ -1 & 2 & -1 & & \\ & \ddots & \ddots & \ddots & \\ & & -1 & 2 & -1 \\ & & & -2 & 2 \end{pmatrix}_{(n+1)\times(n+1)} .$$

The matrix A now is a singular M-matrix since $Ae = 0$, that means
$\mu = 0 \in \sigma(A)$.
Furthermore, it is not hard to see that $\mu_k \in \sigma(A)$ are given by

$$\mu_k = \frac{4}{h^2} \sin^2 \frac{(k-1)\pi}{2n} \, , \tag{5.8}$$

and the corresponding eigenvectors v_k can be chosen as

$$v_k = \sqrt{\frac{2}{1}} \left(1, \cos\frac{(k-1)\pi x_1}{1} , \ldots , \cos\frac{(k-1)\pi x_{n-1}}{1}, (-1)^{k+1} \right)^T , \tag{5.9}$$

for $k=1,\ldots,n+1$.

Now let us analyse some interesting properties of the model problem
under consideration with respect to M-matrices.
First of all, it is obvious that the shapes of the eigenvectors v_k
of the arising irreducible M-matrices A, given by formulas (5.5),
(5.7) and (5.9), are entirely the same as those of the eigenfunctions $u_k(x)$, $x \in \Omega$, of the Problems 5.6.a-c, respectively.
Of course, this analogy does not hold for more general eigenvalue
problems, but it is a striking feature in the simplest Sturm-Liouville

eigenvalue problem considered. The eigenvectors v_k for $k = 1,2,3,4$ are in fact depicted in Figures 5.1.-3. if the interval $\overline{\Omega}$ is subdivided by the uniform grid $\overline{\omega}_h$.

Furthermore, the eigenvalues μ_k of the M-matrices A are less then the corresponding eigenvalues λ_k of the Sturm-Liouville eigenvalue problems. That is

$$\mu_k \leq \lambda_k , \quad k = 1,..,n_s , \tag{5.10}$$

where
$$n_s = \begin{cases} n-1 & \text{in Approximation 5.6.a ,} \\ n & \text{in Approximation 5.6.b ,} \\ n+1 & \text{in Approximation 5.6.c .} \end{cases}$$

In (5.10), equality holds only for $k = 1$ in Approximation 5.6.c, where $\mu_1 = \lambda_1 = 0$.

The point we wish to emphasize here is that in each case the eigenvector v_1, which corresponds to the smallest eigenvalue μ_1 of A has components only of one sign. In our choice, we have $v_1 > 0$, see (5.5), (5.7) and (5.9) for $k = 1$. In this we have got a direct application of Theorem 2.6. and by Corollary 2.7. there cannot exist a second eigenvector of A with all components of one sign.

Thus the considered FDM approximation to Problem 5.6.a-c via eigenvalue problems for irreducible M-matrices A preserves the remarkable Property 5.4.

In Problem 5.6. and its Approximation 5.6. we picked out only three examples of all the possible homogeneous boundary conditions at the boundary $\partial\Omega = \{0,1\}$ for the Sturm-Liouville eigenvalue problem under consideration. If, in other cases, an eigenvalue problem for irreducible M-matrices A also results from the FDM approximation, we also get a reflection of Property 5.4.

After Problem 5.6. we turn to the general one-dimensional Sturm-Liouville eigenvalue problem.

<u>Problem 5.7.</u>

Let $\Omega = (0,1)$, $0 < 1 < \infty$. Consider

$$Lu = - (pu')' + q u = \lambda u , \quad x \in \Omega ,$$

$$\alpha_0 u(0) - \beta_0 u'(0) = 0,$$

$$\alpha_1 u(1) + \beta_1 u'(1) = 0.$$

We assume that $p(x) \in C^1(\Omega)$, $q(x) \in C(\Omega)$ with $p(x) > 0$, $q(x) \geqslant 0$
for $x \in \Omega$ and $\alpha_i \geqslant 0$, $\beta_i \geqslant 0$, $i=0,1$ with $\alpha_i + \beta_i > 0$, $i=0,1$.
For simplicity, we exclude the case $\lambda = 0 \in \mathfrak{S}(L)$, see Property 5.3.
Then the spectrum $\mathfrak{S}(L)$ contains countably many simple positive
eigenvalues, see Property 5.2. That is

$$0 < \lambda_1 < \lambda_2 < \ldots < \lambda_k < \ldots \tag{5.11}$$

and the corresponding eigenfunctions $u_k(x)$, $k=1,2,\ldots$ are L_2-ortho-
gonal, i.e. $(u_k,u_j) = \delta_{kj}$. There is only the eigenfunction $u_1(x)$
of one sign for $x \in \Omega$, which can be chosen $u_1(x) > 0$, $x \in \Omega$, see
Property 5.4.

We shall now consider a three point FDM approximation of Problem 5.7.
on a uniform grid $\overline{\omega}_h$ defined by (5.2).

<u>Approximation 5.7.</u>

Let y_i be the same approximate values as in (5.3). To discretize
$Lu = \lambda u$, we use

$$- \frac{p_{i-1/2}}{h^2} y_{i-1} + \left(\frac{p_{i-1/2} + p_{i+1/2}}{h^2} + q_i \right) y_i - \frac{p_{i+1/2}}{h^2} y_{i+1} = \mu y_i,$$

for $i=1,\ldots,n-1$, putting $p_{i-1/2} = p(x_{i-1/2}) = p((x_{i-1} + x_i)/2)$,
$q_i = q(x_i)$, see $[23]$.

Furthermore, suppose that the two homogeneous boundary conditions
on $\partial\Omega$ of Problem 5.7. are also discretized and then incorporated
into the system of difference equations.

Omitting the details, we write the resulting matrix eigenvalue problem in the form

$$A y = \mu y,$$

$y = (y_0,\ldots,y_n)^T$. Motivated by the three-point difference approximations of Lu, our assumption is now that A is a nonsingular irreducible and tridiagonal M-matrix. It follows from Section 2.2.4. that
the spectrum $\mathfrak{S}(A) \subset R_+^1 \setminus \{0\}$, which also holds in the case $A \neq A^T$.
The eigenvalues of A are all simple, see $[19]$.
We emphasize again that the irreducible M-matrix A has one and only
one eigenvector v_1, whose components are all of one sign and which
therefore can be chosen $v_1 > 0$. This eigenvector v_1 corresponds to
the smallest eigenvalue $\mu_1 \in \mathfrak{S}(A) = \{\mu_i: 0 < \mu_1 < \ldots < \mu_{n+1}\}$.

In this sence Approximation 5.7. reflects the important Property 5.4. of Problem 5.7.

5.3. A finite difference approximation of a higher-dimensional Sturm-Liouville eigenvalue problem

In the present section we try to give an impression of how Property 5.4. carries over from special higher-dimensional Sturm-Liouville eigenvalue problems into its FDM approximation, which involes M-matrices. For this purpose and for brevity of notation, we first choose a two-dimensional example from which a generalization is then easily seen.

<u>Problem 5.8.</u>

Let $\Omega = (0,1)^2 \subset R^2$. Consider for a constant $q \geqslant 0$ the following Sturm-Liouville eigenvalue problem.

$$Lu = -\Delta u + q u = \lambda u, \quad x \in \Omega,$$
$$u = 0, \quad x \in \partial\Omega.$$

The whole solution of Problem 5.8. is well-known, see [11]. Thus, for the eigenvalues we have

$$\lambda_{ks} = \left(\frac{k\pi}{1}\right)^2 + \left(\frac{s\pi}{1}\right)^2 + q \in \sigma(L), \quad (5.12)$$

and the corresponding eigenfunctions can be chosen as

$$u_{ks}(x) = \sin\frac{k\pi x_1}{1} \sin\frac{s\pi x_2}{1}, \quad (5.13)$$

for $k,s = 1,2,\dots$.

It can be seen immediately that $\lambda_{11} > 0$ is the smallest eigenvalue of Problem 5.8. and the corresponding eigenfunction $u_{11}(x)$ is of one sign for $x \in \Omega$, that is, $u_{11}(x) > 0$, $x \in \Omega$. There is no other eigenfunction of Problem 5.8. which exhibits the same property.

Now we consider an FDM approximation to Problem 5.8. on a uniform square grid defined by

$$\overline{\omega}_h = \left\{x_{ij} = (ih,jh),\ 0 \leqslant i,j \leqslant n\right\} \setminus \left\{(0,0),(0,1),(1,0),(1,1)\right\}, (5.14)$$

with step size $h = 1/n$ via the usual five point difference star.

Let $\overline{\omega}_h = \omega_h + \gamma_h$, $\omega_h = \left\{x_{ij} \in \overline{\omega}_h \text{ with } x_{ij} \in \Omega\right\}$,

$\gamma_h = \left\{x_{ij} \in \overline{\omega}_h \text{ with } x_{ij} \in \partial\Omega\right\}$.

Approximation 5.8.

$$(- D^1_+ D^1_- - D^2_+ D^2_- + q\, I)\, y_{ij} = y_{ij}\ , \quad x_{ij} \in \omega_h\ ,$$

$$y_{ij} = 0\ , \quad x_{ij} \in \gamma_h\ .$$

The D^k_+, D^k_- for $k = 1,2$ are difference operators with respect to
the x_k-axis direction, see also Section 4.3.2.

Let $y \in R^{(n-1)^2}$ be a vector which contains the approximate values
y_{ij} for $1 \leq i,j \leq n-1$, listed row by row. Then we rewrite Approximation 5.8. in matrix form, that is

$$A\, y = \mu\, y\ , \tag{5.15}$$

where A is the following block tridiagonal matrix

$$A = \frac{1}{h^2}\begin{pmatrix} A & -I & & & \\ -I & A & -I & & \\ & \ddots & \ddots & \ddots & \\ & & -I & A & -I \\ & & & -I & A \end{pmatrix}_{(n-1)^2 \times (n-1)^2}$$

with $A = \text{tridiag}(-1, 4+h^2 q, -1)$ of order $n-1$.

It is then quite easy to see that $A = A^T$ is an irreducible weakly
diagonal dominant L-matrix if $q = 0$ or it is an irreducible strongly diagonally dominant L-matrix if $q > 0$. Thus, A is a nonsingular
M-matrix in each of the two cases. That means, A is a Stieltjes
matrix.

The whole solution of the matrix eigenvalue problem (5.15) is known
and we shall cite it from [11]. We have

$$\mu_{ks} = \frac{4}{h^2}\left(\sin^2\frac{k\pi}{2n} + \sin^2\frac{s\pi}{2n}\right) + q \in \sigma(A)\ , \tag{5.16}$$

and the corresponding eigenvectors can be chosen as

$$v_{ks} = \left(\sin\frac{k\pi ih}{1}\ \sin\frac{s\pi jh}{1}\right)^{n-1}_{i,j=1}\ , \tag{5.17}$$

for all $1 \leq k,s \leq n-1$.

We see that $\mu_{11} > 0$ is the smallest eigenvalue of A and the corresponding eigenvector v_{11}, chosen here such that $v_{11} > 0$, is the only
one eigenvector of the Stieltjes matrix A which has all components
of one sign. Furthermore, Approximation 5.8. again has the remarkable property that the shapes of the eigenvectors v_{ks} of A may be

the same as the shapes of the eigenfunctions $u_{ks}(x)$, $x \in \Omega$, for $1 \leqslant k,s \leqslant n-1$.

We remark that the eigenvectors (5.17) of the discretized two-dimensional Sturm-Liouville eigenvalue Problem 5.8. over $\Omega = (0,1)^2$ are in fact the tensor product of the eigenvectors v_k and v_s of the discretized one-dimensional analogue over $\Omega = (0,1)$ given by (5.5). This shows how to construct further examples of higher-dimensional Sturm-Liouville eigenvalue problems with explicitly given solutions and which obviously reflect Property 5.4.

6. INVERSE M-MATRICES AND GREEN'S FUNCTIONS

In the present chapter we turn again to discretizations of second order linear elliptic boundary value problems which involve non-singular M-matrices. We shall focus attention on the shape of the inverses of these M-matrices which exhibit behaviour similar to that of Green's functions of the discretized problems. This fact will be illustrated by some examples in which the inverses of the arising nonsingular M-matrices are explicitly known.

Here, we can only show that discretizations of certain elliptic problems involving diagonally dominant M-matrices may reflect some basic properties of Green's functions in their inverses. For this purpose, we need the maximum principle for inverse column entries, see also Section 3.4.3.

Here, we limit our considerations to one-dimensional elliptic problems, since it is well-known that Green's functions are unbounded even for the simplest two- or higher-dimensional second order elliptic problems, see [11,27]. We remark that an estimate for Green's discrete function of the discretized two-dimensional operator $-\Delta$ is given in [11]. This example provides an impression of the difficulties to cope with.

6.1. Two simple examples

We start with the discussion of two simple elliptic problems discretized by FDM approximation to illustrate the purpose of this chapter.

Let $\Omega = (0,1)$ and consider the following problems.

<u>Problem 6.1.</u>

$$- u'' = f(x) , \qquad x \in \Omega ,$$

$$u(0) = u(1) = 0 .$$

We assume that $f(x) \in C(\overline{\Omega})$.

Green's function $G(x,\xi)$, $(x,\xi) \in \Omega^2$ is given by

$$G(x,\xi) = \begin{cases} x (1 - \xi) , & x \leq \xi, \\ \xi (1 - x) , & x > \xi, \end{cases} \tag{6.1}$$

see [6].

Thus, the solution $u(x)$, $x \in \Omega$ of Problem 6.1. is of the form

$$u(x) = \int_0^1 G(x,\xi) \, f(\xi) \, d\xi . \tag{6.2}$$

Over a uniform grid $\overline{\omega}_h = \{ x_i = ih, \ i=0,\dots,n \} = \omega_h + \gamma_h$ with step size $h = 1/n$, we now consider an FDM approximation to Problem 6.1. assuming y_i to be the approximate values to $u(x_i)$, $x_i \in \overline{\omega}_h$.

<u>Approximation 6.1.</u>

$$- D_+ D_- y_i = f_i , \qquad i=1,\dots,n-1 ,$$

$$y_0 = y_n = 0 ,$$

where $f_i = f(x_i)$.

We write the whole system of difference equations in matrix form

$$A \, y = f , \tag{6.3}$$

where $y = (y_1,\dots,y_{n-1})^T$, $f = (f_1,\dots,f_{n-1})^T$ and

$$A = \frac{1}{h^2} \text{tridiag}(-1,2,-1) .$$

The tridiagonal matrix A is a nonsingular M-matrix. Since A is irreducible, it follows that $A^{-1} > 0$, see Property 3.7.

Exept for a positive factor, its inverse A^{-1} is known from Example 3.34. Let $A^{-1} = (a^-_{ij})$, then we have

$$a^-_{ij} = \begin{cases} \dfrac{h^2}{n}(n-i)j , & j < i , \\[2ex] \dfrac{h^2}{n}(n-j)i , & j \geq i . \end{cases} \tag{6.4}$$

Let us now compare A^{-1} with Green's function (6.1) via (6.2).
For this purpose, we discretize the square $\Omega^2 = (0,1)^2$ by the
uniform square grid ω_h^2,

$$\omega_h^2 = \omega_h \times \omega_h = \left\{ (x_i, \xi_j) : x_i = ih, \ \xi_j = jh, \ i,j=1,\ldots,n-1 \right\}.$$

The remarkable feature is now that

$$\bar{a}_{ij} = h\, G(x_i, \xi_j) \quad \text{for all} \ (x_i, \xi_j) \in \omega_h^2 \qquad (6.5)$$

obviously holds. This fact is illustrated graphically in Figure 6.1.

Fig.6.1.

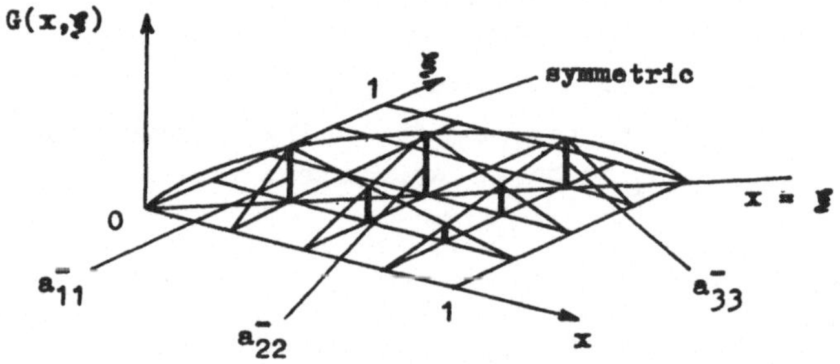

We shall briefly sketch another relationship between A^{-1} and $G(x,\xi)$.
Using (6.2) and the grid $\bar{\omega}_h$, we see that

$$u(x) = \sum_{i=1}^{n} \int_{x_{i-1}}^{x_i} G(x,\xi)\, f(\xi)\, d\xi. \qquad (6.6)$$

Applying now one of the simplest quadrature formulas, that is

$$\int_{x_{i-1}}^{x_i} g(\xi)\, d\xi \approx (x_i - x_{i-1})\, g(x_i) \, ,$$

we derive the following approximate equations to (6.6)

$$y_j = \sum_{i=1}^{n} h\, G(x_j, \xi_i)\, f_i \, , \quad j=1,\ldots,n-1 \, ,$$

which are entirely equivalent to $y = A^{-1}f$.

<u>Problem 6.2.</u>

$$- u'' = f(x) , \qquad x \in \Omega ,$$
$$u'(0) = u(1) = 0 .$$

We again assume that $f(x) \in C(\bar{\Omega})$.

Green's function of Problem 6.2. is then

$$G(x,\xi) = \begin{cases} 1 - x , & x \leq \xi , \\ 1 - \xi , & x > \xi , \end{cases} \qquad (6.7)$$

see [6], and

$$u(x) = \int_0^1 G(x,\xi) \, f(\xi) \, d\xi \qquad (6.8)$$

is the unique solution of Problem 6.2.

<u>Approximation 6.2.</u>

$$y_0 = y_1 ,$$
$$- D_+ D_- y_i = f_i , \qquad i=1,\ldots,n-1 ,$$
$$y_n = 0 .$$

We introduce the matrix form of the difference equation system

$$A \, y = f , \qquad (6.9)$$

where y and f are the same vectors as in (6.3), but

$$A = \frac{1}{h^2} \begin{pmatrix} 1 & -1 & & & \\ -1 & 2 & -1 & & \\ & & \ddots & & \\ & & -1 & 2 & -1 \\ & & & -1 & 2 \end{pmatrix}_{(n-1) \times (n-1)} \qquad (6.10)$$

From Example 3.35., we know the inverse $A^{-1} = (a_{ij}^-)$ of the nonsingular M-matrix (6.10). We find

$$a_{ij}^- = \begin{cases} h^2 \, (n-j) , & j < i , \\ h^2 \, (n-i) , & j \geq i . \end{cases} \qquad (6.11)$$

We see from this

$$a_{ij}^- = h \, G(x_i, \xi_j) \qquad \text{for all} \quad (x_i, \xi_j) \in \omega_h^2 . \qquad (6.12)$$

An illustration is given in Figure 6.2.

Fig.6.2.

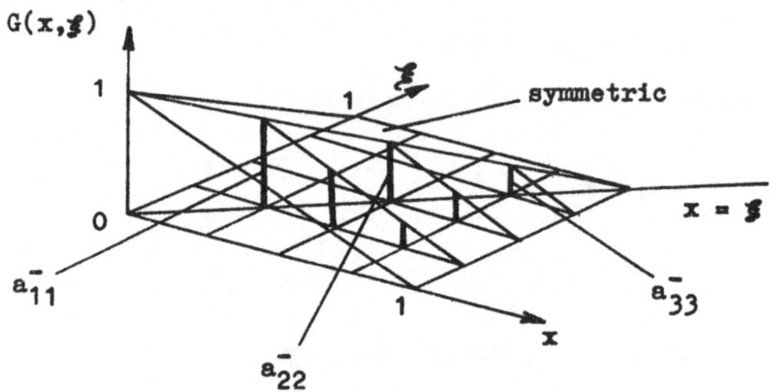

Obviously, the striking feature in both of the examples is that the
equalities given by (6.5) and (6.12) hold. Thus, $G(x,\xi)$ and A^{-1} have
in fact one and the same shape. But this observation is certainly
not the rule in the discretization of more general problems by FDM
or FEM, where nonconstant coefficients occur in the second order
elliptic differential equations.

6.2. Some general analogies between Green's functions and inverse M-matrices

Let us now assess in more detail some observations drawn from the
two examples considered in Section 6.1. with respect to the shape
of Green's function $G(x,\xi)$ and that of A^{-1}, supposing A is a nonsingular M-matrix arising from the discretization of second order
elliptic boundary value problems for which $G(x,\xi)$ is Green's
function.

To be more precise, we consider the following problem.

<u>Problem 6.3.</u>   Let $\Omega = (0,1)$.

$$Lu = \sum_{k=0}^{2} p_k(x)\, u^{(k)} = f(x)\,, \qquad x \in \Omega\,,$$

$$\alpha_0\, u(0) - \beta_0\, u'(0) = u_0\,,$$

$$\alpha_1\, u(1) + \beta_1\, u'(1) = u_1\,.$$

132

Suppose that $p_k(x) \in C(\overline{\Omega})$ for $k = 0,1,2$, $p_2(x) \neq 0$, $x \in \overline{\Omega}$ and
assume that Problem 6.3. has a unique solution $u(x)$, which is re-
presented of the form (6.2), where $G(x,\xi)$ is Green's function of
the problem.

A description of the construction of Green's functions for general
one-dimensional boundary value problems of order $k \geqslant 2$ may be found
in [6,16]. Furthermore, we remark that the number of boundary value
problems, for which the corresponding function of Green's is explicitly known by a formula, is rather limited. But there are several
well-known sufficient conditions for boundary value problems, which
then guarantee special properties of Green's functions.

Suppose now that Problem 6.3. is discretized by FDM or FEM over a
grid $\overline{\omega}_h \subset \overline{\Omega}$ or a subdivision of $\overline{\Omega}$ with nodal points $\overline{\omega}_h$,
respectively.
Thus we get

Approximation 6.3.

Let the linear equation system

$$A\, y \; = \; f \; ,$$

be a discrete approximation to Problem 6.3., where $A = (a_{ij})$,
$y = (y_1,\ldots,y_n)^T$, $f = (f_1,\ldots,f_n)^T$.
Suppose that A is a nonsingular M-matrix.

In the following we shall discuss some desirable properties of A
which need to be generated by the applied discretization method de-
pending on significant properties of Green's function.

First of all, if it is known a priori that $G(x,\xi) > 0$ for $(x,\xi) \in \Omega^2$,
then the nonsingular M-matrix A must be irreducible. Under this condition, we also have $A^{-1} > 0$, see Property 3.7. On the contrary, if
A is reducible, then $A^{-1} \geqslant 0$ possesses a zero pattern which does not
coincide with $G(x,\xi) > 0$ for $(x,\xi) \in \Omega^2$.

Henceforth, let $G(x,\xi) > 0$ for all $(x,\xi) \in \Omega^2$.

Second, if Green's function of Problem 6.3. is symmetric, i.e.,

$$G(x,\xi) \; = \; G(\xi,x) \quad \text{for all} \quad (x,\xi) \in \Omega^2 \; ,$$

then A should be a symmetric nonsingular M-matrix, that is, a
Stieltjes matrix. Thus, A^{-1} is also symmetric in analogy to $G(x,\xi)$.
It is a well-known fact that $G(x,\xi)$ is symmetric if Problem 6.3.
is self-adjoint and positive definite, see [6].

Suppose next that

$$\max_{x \in \Omega} G(x,\xi) \quad = \quad G(\xi,\xi) \ .$$

If $G(\xi,\xi) > G(x,\xi)$ for all $x \neq \xi$ and all $\xi \in \Omega$, then A^{-1} should be strongly diagonally dominant of its column entries. Such an example is illustrated in Problem and Approximation 6.1. A sufficient condition for a nonsingular M-matrix A with an inverse A^{-1} which is strongly diagonally dominant of its column entries is given in Theorem 3.38. But the stated assumption $B > 0$ in the M-matrix representation $A = sI - B$ is not fulfilled by sparse matrices A, which arise in the application of discretization methods. Therefore, one has to look for other results or studies which are not available at the moment.

If $G(\xi,\xi) \geq G(x,\xi)$ for all $x \neq \xi$ and all $\xi \in \Omega$, which includes the latter case, then A^{-1} should be at least weakly diagonally dominant of its column entries. An illustration is given in Problem and Approximation 6.2. Under this assumption, we can directly apply the maximum principle for inverse column entries, see Section 3.4.3. By Theorem 3.32., for this maximum principle it is necessary and sufficient that the nonsingular M-matrix A satisfies $Ae \geq 0, \neq 0$.

REFERENCES

Books

[1] Berg,L.
Lineare Gleichungssysteme mit Bandstruktur. VEB Deutscher
Verlag der Wissenschaften, Berlin, 1986.

[2] Berman,A., Plemmons,R.J.
Nonnegative matrices in the mathematical sciences. Academic
Press, New York, 1979.

[3] Bohl,E.
Finite Modelle gewöhnlicher Randwertaufgaben. B.G. Teubner,
Stuttgart, 1981.

[4] Ciarlet,Ph.
The finite element method for elliptic problems. North-Holland
Publishing Company, Amsterdam, New York, Oxford, 1978.

[5] Collatz,L.
Funktionalanalysis und Numerische Mathematik. Springer-Verlag,
Berlin, Göttingen, Heidelberg, 1968.

[6] Collatz,L.
Eigenwertaufgaben mit technischen Anwendungen. Akademische
Verlagsgesellschaft, Geest & Portig K.-G., Leipzig, 1963.

[7] Doolan,E.P., Miller,J.J.H., Schilders,W.H.A.
Uniform numerical methods for problems with initial and boundary layers, Boole Press, Dublin, 1980.

[8] Fichtenholz,G.M.
Differential- und Integralrechnung III, VEB Deutscher Verlag
der Wissenschaften, Berlin, 1964.

[9] Fiedler,M.
Special matrices and their application in numerical mathe-
matics, Martinus Nijhoff Publishers, Dordrecht, Boston,
Lancester, 1986.

[10] Gantmacher,F.R.
Matrizenrechnung I,II . VEB Deutscher Verlag der Wissenschaften
Berlin, 1970, 1971.

[11] Hackbusch,W.
Theorie und Numerik elliptischer Differentialgleichungen,
B.G. Teubner, Stuttgart, 1986.

[12] Heinrich,B.
Finite difference methods on irregular networks, Akademie-
Verlag, Berlin, 1987.

[13] Ikeda,T.
Maximum principle in finite element models for convection-
diffusion phenomena, North-Holland Publishing Company, Amsterdam, New York, Oxford, 1983.

[14] Il'in,V.P., Kuznezov,Ju.I.
Tridiagonal matrices and their application. Nauka, Moskva, 1985
(Russian).

[15] Kačur,J.
Method of Rothe in evolution equations. Teubner-Texte zur
Mathematik, Vol. 80, B.G.Teubner, Leipzig, 1985.

[16] Kamke,E.
Differentialgleichungen, Lösungsmethoden und Lösungen – Gewöhn-
liche Differentialgleichungen, Akademische Verlagsgesellschaft,
Leipzig, 1959.

[17] Lambert,J.D.
Computational methods in ordinary differential equations.
John Wiley & Sons, London, New York, Sydney, Toronto, 1973.

[18] Maeß,G.
Vorlesungen über Numerische Mathematik, I, II, Akademie-Verlag,
Berlin, 1984, 1988.

[19] Marcus,M., Minc,H.
A survey of matrix theory and matrix inequalities, Allyn and
Bacon, Inc., Boston, 1964.

[20] Ortega,J.M., Poole,W.G.
An introduction to numerical methods for differential equations.
Pitman Publishing Inc., 1981.

[21] Ortega,J.M., Rheinboldt,W.C.
Iterative solution of nonlinear equations in several variables.
Academic Press, New York, London, 1970.

[22] Protter,M.H., Weinberger,H.F.
Maximum principles in differential equations. Prentice-Hall,
Englewood Cliffs, 1967.

[23] Samarskij,A.A.
Theorie der Differenzenverfahren, Akademische Verlagsgesellschaft, Geest & Portig K.-G., Leipzig, 1984.

[24] Smoller,J.
Shock waves and reaction-diffusion equations. Springer-Verlag,
New York, Heidelberg, Berlin, 1983.

[25] Strang,G., Fix,G.
An analysis of the finite element method. Prentice-Hall, Inc.,
Englewood Cliffs, N.J., 1973.

[26] Varga,R.S.
Matrix iterative analysis, Prentice-Hall, Englewood Cliffs,
1962.

[27] Vladimirov,V.S.
Equations of mathematical physics, Nauka, Moskva, 1971,
(Russian).

[28] Voevodin,V.V., Kuznezov,Ju.A.
Matrices and computation, Nauka, Moskva, 1984, (Russian).

[29] Young,D.
Iterative solution of large linear systems, Academic Press,
New York, 1971.

Publications

[30] Alefeld,G., Schneider,N.
On square roots of M-matrices,
Lin.Alg.Appl., 42, 1982, 119 - 132.

[31] Ando,T.
Inequalities for M-matrices,
Linear and Multilinear Algebra, 8, 1980, 291 - 316.

[32] Csordas,G., Varga,R.S.
Comparations of regular splittings of matrices,
Num.Math., 44,1984, 23 - 35.

[33] Farrell,A.P.
Sufficient conditions for the uniform convergence of a diffe-
rence scheme for a singularly perturbed turning point problem.
SIAM J. Num. Anal., Vol.25, Nr. 3, 1988, 618 - 643.

[34] Fiedler,M., Pták,V.
On matrices with non-positive off-diagonal elements and po-
sitive principal minors,
Czechoslovak Math. J., 12, 1962, 382 - 400.

[35] Fiedler,M., Pták,V.
Diagonally dominant matrices.
Czechoslovak Math. J., 17, 1967, 420 - 433.

[36] Fiedler,M., Schneider,H.,
Analytic functions of M-matrices and generalizations.
Linear and Multilinear Algebra, 13, 1983, 185 - 201.

[37] Hemker,P.N.
Numerical aspects of singular perturbation problems,
Math. Centrum, Amsterdam, NW 133/82.

[38] Il'in,A.M.
Differencing scheme for a differential equation with a small
parameter affecting the highest derivative,
Math. Notes, 6, 1969, 596 - 602.

[39] Johnson,Ch.,R.
Closure properties of certain positivity class of matrices
under various algebraic operations,
Lin.Alg.Appl., 97, 1987, 243 - 247.

[40] Johnson,Ch.,R.
Inverse M-matrices.
Lin.Alg.Appl., 47, 1982, 195 - 216.

[41] Lorenz,J.
Zur Inversemonotonie diskreter Probleme,
Num.Math., 27, 1977, 227 - 238.

[42] Minkowski,H.
Zur Theorie der Einheiten in den algebraischen Zahlenkörpern,
Nachr.K.Ges.Wiss.Gött., Math.-Physik. Klasse, 1900, 90 - 93.
Gesammelte Abhandlungen von H.Minkowski, 1. Band, B.G.Teubner,
Leipzig und Berlin, 1911.

[43] Micchelli,Ch.A., Willoughby,R.A.
On functions which preserve the class of Stieltjes matrices.
Lin.Alg.Appl., 23, 1979, 141 - 156.

[44] Ostrowski,A.
Über die Determinanten mit überwiegender Hauptdiagonale.
Comment. Math. Helv., 10, 1937, 69 - 96.

[45] Plemmons,R.J.
M-matrix characterizations I - nonsingular M-matrices.
Lin.Alg.Appl., 18, 1977, 175 - 188.

[46] Poole,G., Boullion,T.
A survey on M-matrices.
SIAM Rev., 16, 1974, 419 - 427.

[47] Sierksma,G.
Non-negative matrices: The open Leontief model.
Lin.Alg.Appl., 26, 1979, 175 - 201.

[48] Stadelmaier,M.W., Rose,N.J., Poole,G.D., Meyer,C.D.J.
Nonnegative matrices with power invariant zero patterns.
Lin.Alg.Appl., 42, 1982, 23 - 29.

[49] Stoyan,G.
On a maximum principle for matrices and on conservation of
monotonicity with applications to discretization methods.
ZAMM, 62, 1982, 375 - 381.

[50] Stoyan,G.
On maximum principles for monotone matrices.
Lin.Alg.Appl., 78, 1986, 147 - 161.

[51] Varga,R.S., Cai,D.-Y.
On the LU factorization of M-matrices,
Num.Math., 38, 1981, 179 - 192.

[52] Willoughby,R.A.
The inverse M-matrix problem.
Lin.Alg.Appl., 18, 1977, 75 - 94.

[53] Windisch,G.
A maximum principle for systems with diagonally dominant
M-matrices.
In "Mathematical Research", Vol. 36: Discretization in Diffe-
rential Equations and Enclosures. Akademie-Verlag, Berlin,
1987, 243 - 250.

G. Schaar / M. Sonntag / H.-M. Teichert

Hamiltonian Properties of Products of Graphs and Digraphs

This book gives a survey on the main results concerning the subject described by the title, also considering the contributions made by the authors in this field. The central object is to study the dependence of the Hamiltonian behaviour of given products of graphs on properties of the factors. Moreover, the classical products (Cartesian sum, lexicographic product, disjunction, Cartesian product, normal product) are particularly investigated in connection with such Hamiltonian properties as traceability, Hamiltonicity, higher Hamiltonicity, Hamiltonian connectedness, strong path-connectedness, pancyclicity, decomposability into Hamiltonian cycles. The parallel treatment of this set of problems for undirected and directed graphs provides the possibility of a comparative consideration with regard to similarities and differences.

Bd. 108, 148 S., 1988, DDR 15,50 M; Ausland 15,50 DM,
ISBN 3-322-00501-1

Seminar Analysis of the Karl-Weierstraß-Institute 1986/87

Ed. by B.-W. Schulze and H. Triebel

The Teubner-Text 'Seminar Analysis' is the continuation of a corresponding series published by the Karl-Weierstraß-Institute of Mathematics of the Academy of Sciences of the GDR 1981 - 1985. The volume 1985/86 appeared as the Teubner-Text 96.
The main aim of this series is the publication of survey papers on modern analysis, in particular functional analytic and structure methods in partial differential equations, complex function theory, mathematical physics, global analysis and differential geometry.
Another part contains short announcements of outstanding results on these subjects.
The present volume contains articles on elliptic operators on non-compact manifolds, in particular with conical singularities, global analysis, infinite-dimensional supermanifolds, functional analysis, propagation of singularities.

Bd. 106, 332 S., 1988, DDR 34,50 M, Ausland 34,50 DM,
ISBN 3-322-00503-8

C. A. Kalužnin / P. M. Beleckij / V. Z. Fejnberg

Kranzprodukte

Das vorliegende Buch behandelt Kranzprodukte von Permutationsgruppen (und Transformationsgruppen) und unterscheidet sich damit von vielen anderen Publikationen, in denen Kranzprodukte anderer algebraischer Strukturen (z.B. abstrakte Gruppen oder Halbgruppen) betrachtet werden. Kranzprodukte von Permutationsgruppen wurden unter dem Namen "produit complêt" von L. A. Kaloujnine (= Kalužnin) und M. I. Krasner in den 40er Jahren eingeführt. Historisch kann man den Begriff des Kranzproduktes bis in das 19. Jahrhundert zurückverfolgen.
Anwendungen des Kranzproduktes für Permutationsgruppen gibt es in der mathematischen Chemie und der Informatik. Innerhalb der Mathematik finden Kranzprodukte Anwendungen besonders in der abstrakten Gruppentheorie wie auch in der Theorie der Permutationsgruppen und führen zu wichtigen Ergebnissen in beiden Theorien (Schreiersches Gruppenerweiterungsproblem, Geometrie ultrametrischer Räume).
Das Buch wendet sich vor allem an Studenten und Hochschullehrer auf dem Gebiet der reinen und angewandten Mathematik, besonders der Informatik, aber auch an Mathematiker und Informatiker mit algebraischen Interessen.

Bd. 101, 167 S., DDR 17,50 M, Ausland 17,50 DM, ISBN 3-322-00425-2

Diese Reihe wurde geschaffen, um eine schnellere Veröffentlichung
mathematischer Forschungsergebnisse und eine weitere Verbreitung von
mathematischen Spezialvorlesungen zu erreichen. TEUBNER-TEXTE werden
in deutsch, englisch, russisch oder französisch erscheinen. Um Aktu-
alität der Reihe zu erhalten, werden die TEUBNER-TEXTE im Manuskript-
druck hergestellt, da so die geringeren drucktechnischen Ansprüche
eine raschere Herstellung ermöglichen. Autoren von TEUBNER-TEXTEN
liefern an den Verlag ein reproduktionsfähiges Manuskript. Nähere
Auskünfte darüber erhalten die Autoren vom Verlag.

This series has been initiated with a view to quicker publication of
the results of mathematical research-work and a widespread circula-
tion of special lectures on mathematics. TEUBNER-TEXTE will be published in German, English, Russian or French. In order to keep this
series constantly up to date and to assure a quick distribution, the
copies of these texts are produced by a photographic process (small-
offset printing) because its technical simplicity is ideally suited
to this type of publication. Authors supply the publishers with a
manuscript ready for reproduction in accordance with the latter's
instructions.

Cette série des textes a été créée pour obtenir une publication plus
rapide de résultats de recherches mathématiques et de conférences
sur des problèmes mathématiques spéciaux. Les TEUBNER-TEXTE seront
publiés en langues allemande, anglaise, russe ou française. L'actua-
lité des TEUBNER-TEXTE est assurée par un procédé photographique
(impression offset). Les auteurs des TEUBNER-TEXTE sont priés de
fournir à notre maison d'édition un manuscrit prêt à être reproduit.
Des renseignements plus précis sur la form du manuscrit leur sont
donnés par notre maison.

Эта серия была создана для обеспечения более быстрого публикования
результатов математических исследований и более широкого распростра-
нения математических лекций на специальные темы. Издания серии
ТОЙБНЕР-ТЕКСТЕ будут публиковаться на немецком, английском, русском
или французском языках. Для обеспечения актуальности серии, ее издания будут изготовляться фотомеханическом способом. Таким образом,
более скромные требования к полиграфическому оформлению обеспечат
более быстрое появление в свет. Авторы изданий серии ТОЙБНЕР-ТЕКСТЕ
будут предоставлять издательству рукописи, удовлетворящие требова-
ниям фотомеханического печатания. Более подробные сведения авторы
получат от издательства.

BSB B. G. Teubner Verlagsgesellschaft, Leipzig
DDR - 7010 Leipzig, Postfach 930